BrightRED Study Guide

Curriculum for Excellence

N5

MATHEMATICS

Brian J Logan

BrightRED
PUBLISHING

First published in 2013 by:
Bright Red Publishing Ltd
1 Torphichen Street
Edinburgh
EH3 8HX

A CIP record for this book is available from the British Library

ISBN 978-1-906736-41-5

With thanks to:
PDQ Digital Media Solutions Ltd (layout), Ivor Normand (copy-edit and proof-read)

Cover design by Caleb Rutherford – e i d e t i c

Acknowledgements
Every effort has been made to seek all copyright holders. If any have been overlooked, then Bright Red Publishing will be delighted to make the necessary arrangements.

Alistair Michael Thomas/Shutterstock.com (p 8); Africa Studio/Shutterstock.com (p 40); Natursports/Shutterstock.com (p 61); Monkey Business Images/Shutterstock.com (p 64); Caleb Rutherford (p 65); Tihis/Dreamstime.com (p 72); Joe Belanger/Shutterstock.com (p 75); planet5D LLC/Shutterstock.com (p 77); Susan Trigg/istockphoto (p 80); Palto/Shutterstock.com (p 91); Miroslav K/Shutterstock.com (p 92); Microstock Man/Shutterstock.com (p 102); 36clicks/Dreamstime.com (p 109); vichie81/Shutterstock.com (p 114); Adisa/Shutterstock.com (p 116); ARENA Creative/Shutterstock.com (p 118); Verdateo/Shutterstock.com (p 125); hypnotype/Shutterstock.com (p 126).

Printed and bound in the UK by Martins the Printers.

CONTENTS

INTRODUCING NATIONAL 5 MATHEMATICS

1 EXPRESSIONS AND FORMULAE

2 RELATIONSHIPS

3 APPLICATIONS

A-Z GLOSSARY

INTRODUCING NATIONAL 5 MATHEMATICS

PREFACE

Welcome to this study guide for National 5 Mathematics. The fact that you are reading this is evidence of two very important facts. The first fact is that someone believes that you are capable of passing the course assessment. It may be a teacher or lecturer who has that belief, but the most important person to have such belief is you yourself. The second fact is that by reading this guide, you are proving your determination to pass the course assessment.

In order to be successful at National 5 Mathematics, you will have to prepare properly. Imagine someone who is preparing to sit their driving test. Before the test, that person would have taken an adequate number of lessons, studied the Highway Code, practised for the theory test and taken advice from their instructor on when they were ready to pass. There are many similarities in preparing to pass assessments in school or college. You must attend lessons, study the course, practise key examples and ask for advice about areas of concern. This guide is designed to help you to achieve your aims, but it is not enough on its own. You can ask your teacher or lecturer for advice, discuss the course with friends and possibly attend study classes.

HOW TO USE THIS BOOK

The book starts with a Key Revision section followed by three chapters covering the National 5 course. These chapters cover the three units of the course: Expressions and Formulae, Relationships, and Applications. There are 60 two-page topic spreads covering the major aspects of National 5 Mathematics. It is possible that you will be taught topics in an order different from this book, although this is the order in which the units appear on the Scottish Qualifications Authority (SQA) website.

Each spread contains the key elements from each topic. There are 'Don't Forget' hints which include vital areas to focus on within each spread.

Each spread contains advice, formulae, diagrams and examples of exam standard with solutions and hints for every topic.

Each topic ends with a section called 'Things to do and think about' containing examples for you to practise. There will be an online link containing the solution to all these examples.

Another innovation in the book is the BrightRED Digital Zone, an online source of examples and solutions covering the entire syllabus. The examples will range from easy to medium up to hard and exam style. Solutions will be provided to all examples. This should produce a useful resource as you study throughout the session.

Also, there are suggested video and online links with each spread. These will range across YouTube clips of teachers in action, visual guides, online calculators and even the occasional song. Sometimes it is difficult to explain something adequately in words, whereas it is simpler to explain visually. It is hoped that the videos may assist you with topics of this type.

As you work through the book, try the examples *before* you read the solutions if possible. Remember too that this book on its own is not enough to ensure that you are successful at National 5 Mathematics. Success also requires regular attendance at class, hard work and proper preparation. If there is an area of the syllabus that you feel unsure of, then you should address the situation and ask for appropriate help.

Do not leave things to the last minute when studying, and do not try to do too much at the one time. The way to get the maximum benefit from the guide is to revise regularly in fairly small doses. This requires planning, so you will need to be organised.

ONLINE

This book is supported by the Bright Red Digital Zone. Visit www.brightredbooks.net/N5Maths to log on!

ASSESSMENT

During the course of the session, you will have to complete three unit assessments, one on each area of the syllabus. The three areas are Expressions and Formulae, Relationships, and Applications. You must pass all three assessments. You will be given the opportunity to re-sit these assessments if required.

The final key part of the assessment will be the external exam. This will consist of two question papers. Paper 1 will be a non-calculator paper worth 40 marks. In Paper 2, you will be allowed to use a calculator. Paper 2 is worth 50 marks. This final external exam will be the equivalent of your cup final. As such you must have prepared thoroughly for the occasion so that you peak at the right time. Part of this will involve studying likely exam scenarios by reading sample papers and questions in advance.

CALCULATORS

A scientific calculator is perfectly suitable for this course. It is fine to use a more advanced calculator such as the type which has the facility to draw graphs. Become familiar with your own calculator and learn to use it accurately and consistently at home and in school or college.

Finally on calculators, we will look at a few useful keys that can help you as you work through the course. Firstly, make sure your calculator is in degree mode for use in questions on trigonometry. As a check, you should key in tan 45. If you get 1, you are in degree mode. If not, make sure you know how to return to degree mode. Secondly, when doing calculations involving a circle, always use the π key for accuracy instead of 3·14, which is approximate. Thirdly, there are two useful keys for squaring numbers (x^2) and for calculating powers (y^x) or ($\wedge$). These keys will be very useful. Check that you can calculate that $13^2 = 169$ and $5^4 = 625$ as examples.

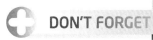

DON'T FORGET

There are many keys on a calculator that you will never use at the level of National 5 Mathematics. If you are in any doubt about this, ask your teacher or lecturer for advice.

ADVICE ON STUDYING

It is essential that regular practice should form part of your studying process. This is true of any venture in life where you wish to become really proficient, whether it is mathematics, sport, driving, cookery or anything else. A famous golfer had a routine to practise putting in which he placed his golf ball 3 feet from the hole and practised holing the putt 100 times in a row. If he missed at any time, he would start at 1 again, even if he missed putt number 99. He did that so that when it mattered, in a competition, he would be so good that the chances of missing under pressure would be unlikely. He showed dedication to becoming a champion. No-one is suggesting that you need to go to such extremes to become good at mathematics; but regular practice will undoubtedly help you to improve your skills as well as giving you confidence that you are heading in the right direction.

So, good luck and good studying!

ONLINE

The SQA website may help with your studying, as there will be information about the course as well as practice material.

KEY REVISION

MEASUREMENT

As you prepare for National 5 Mathematics, you should be aware of many important topics from National 4 Mathematics. During the sections to follow, many reminders will be given to make it clear what you should know. We shall start here with a reminder of some key facts on measurement.

Length: 1 cm = 10 mm; 1 m = 100 cm; 1 km = 1000 m
Volume: 1 litre (1 l) = 1000 ml; 1 ml = 1 cm³
Mass: 1 kg = 1000 g

Here is a list of formulae you will be expected to have memorised:

Area of a rectangle: $A = lb$; Area of a square: $A = l^2$; Area of a triangle: $\frac{1}{2}bh$;
Volume of a cuboid: $V = lbh$; Volume of a cube: $V = l^3$; Volume of a prism: $V = Ah$

You must also memorise two key formulae concerning the circle:

Circumference of a circle: $C = \pi d$; Area of a circle: $A = \pi r^2$

EXAMPLE:

The window in a church is in the shape of a rectangle with a semi-circular top, as shown in this diagram.

Calculate the area of the window, giving your answer to the nearest tenth of a square metre.

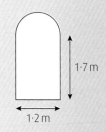

1·7 m

1·2 m

SOLUTION:

Area of rectangle: $A = lb = 1{\cdot}7 \times 1{\cdot}2 = 2{\cdot}04$ m²
Area of semi-circle = $\frac{1}{2} \pi r^2 = 0{\cdot}5 \times \pi \times 0{\cdot}6^2 = 0{\cdot}57$ m²
Area of window = 2·04 + 0·57 = 2·61 m²

Hence area of window = 2·6 m² (to the nearest tenth of a m²).

NOTE: Did you remember to halve the diameter (1·2) to get the radius (0·6)?

PERCENTAGES

You should be able to carry out basic percentage calculations with and without a calculator. Suppose, for example, that you were asked to calculate 35% of £140. Without a calculator, you could find 10% of £140 = £14. If you multiply by 3, you find that 30% of £140 = £42. Then notice that 5% of £140 = £7 (half of 10%), leading to the solution that 35% of £140 = £49. With a calculator, calculate $\frac{35}{100} \times 140$ or $0{\cdot}35 \times 140$, leading to £49.

Remember also such well-known percentages as 20% = $\frac{1}{5}$, 25% = $\frac{1}{4}$, $33\frac{1}{3}$% = $\frac{1}{3}$, $66\frac{2}{3}$% = $\frac{2}{3}$ and 75% = $\frac{3}{4}$.

EXAMPLE:

A bicycle cost £250 new. It was later sold for £230. Express the loss as a percentage of the cost price.

SOLUTION:

Actual loss = £250 − £230 = £20
Loss as percentage of cost price = $\frac{20}{250} \times 100 = 8\%$.

THE THEOREM OF PYTHAGORAS

Theorem of Pythagoras:

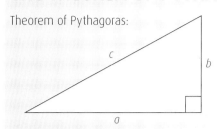

$a^2 + b^2 = c^2$

EXAMPLE:

In a right-angled triangle, the two shorter sides are 8 centimetres and 15 centimetres in length. Find the length of the hypotenuse, the longest side in the triangle.

SOLUTION:

$c^2 = a^2 + b^2 = 8^2 + 15^2 = 64 + 225 = 289$

Hence $c = \sqrt{289} = 17$, therefore the length of the hypotenuse is 17 cm.

 DON'T FORGET

The Theorem of Pythagoras is a key piece of mathematics which will feature many times in the National 5 course, so you must be able to use it accurately.

TRIGONOMETRY

We will study trigonometry in great depth in the National 5 course. Before that, we have revision of two basic trigonometric calculations.

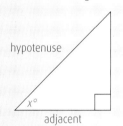

$$\tan x° = \frac{\text{opposite}}{\text{adjacent}}$$

$$\sin x° = \frac{\text{opposite}}{\text{hypotenuse}}$$

$$\cos x° = \frac{\text{adjacent}}{\text{hypotenuse}}$$

EXAMPLE:

In the diagram, find the length of the side marked x.

SOLUTION:

$\sin 38° = \frac{x}{16} \Rightarrow x = 16 \times \sin 38° = 9.9$ cm

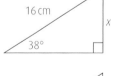

EXAMPLE:

In the diagram, find the size of the angle marked x.

SOLUTION:

$\cos x° = \frac{16}{25} = 0.64 \Rightarrow x = \cos^{-1} 0.64 = 50°$

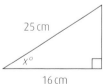

There are many other topics which you should be familiar with before you start the National 5 course. We do not have time to go over them all here. They include fractions, rounding, skills in algebra such as solving simple equations, facts concerning angles, properties of quadrilaterals, some basic statistics as well as graphs and charts. Where such skills occur, reminders will be given throughout the sections to follow.

 THINGS TO DO AND THINK ABOUT

Find the volume of the triangular prism shown here.

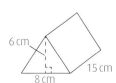

SIGNIFICANT FIGURES

APPROXIMATION

In real-life situations, numbers are often rounded to approximate numbers depending on the accuracy required. For example, we might measure our height to the nearest centimetre; the interest given by a bank may be given to the nearest penny; the population of the United Kingdom according to the 2011 census was 63 182 000 to the nearest thousand. It would equally be appropriate to say that the population was around 63 million.

The process of giving approximate answers involves rounding. We shall look at different ways of rounding in this section, including significant figures.

EXAMPLE:

An example on rounding you should be familiar with concerns foreign currency. Suppose you return from a holiday in Spain with 100 euros and wish to change this money into pounds sterling. You find that the exchange rate is £1 = 1·174 euros. Convert the euros to pounds sterling, giving your answer to the nearest penny.

SOLUTION:

100 euros = 100 ÷ 1·174 pounds = £85·178 875 64 = £85·18 (to the nearest penny)

DECIMAL PLACES

In the previous example, we were effectively rounding the answer to two decimal places, as we required two numbers after the decimal point in the solution for the pence. Using decimal places (the number of figures that appear after the decimal point) is a common way of rounding.

For example, 6·753 871 has six decimal places. Suppose you were asked to round it to three decimal places.

- 6·753 871 lies between 6·753 and 6·754

- look at the figure in the fourth decimal place (6·753 **8**) – that is, 8

- since this figure is **more than 5**, round up to 6·754

Check that 6·753 871 rounded to two decimal places would be 6·75.

EXAMPLE:

The radius of a circle is 15 centimetres. Calculate its circumference. Give your answer correct to one decimal place.

SOLUTION:

$C = \pi d = \pi \times 30 = 94\cdot247\,779\,61 = 94\cdot2\,cm$ (to one decimal place)

SIGNIFICANT FIGURES

Another method of rounding to an approximate number is to use significant figures (sig. figs for short). In a number, all figures are significant *except* zeros that are used simply to indicate the position of the decimal point. However, zeros in between other significant figures are themselves significant.

Hence, 589 346 has 6 sig. figs; 715·28 has 5 sig. figs; 4045 has 4 sig. figs. Note that 0·003 48 has only 3 sig. figs, as the zeros at the start indicate the position of the decimal point, whereas 0·003 048 has 4 sig. figs, as the zero between the 3 and 4 is significant. A measurement such as 25·0 centimetres has 3 sig. figs, as the final zero tells you that it is a more accurate measurement than simply 25 centimetres.

Be careful with whole numbers. A crowd at a football match of 25 000 to the nearest thousand only has 2 sig. figs. However, when you say that there are 90° in a right angle, there are 2 sig. figs, as the number is exactly 90. In other words, when there are trailing zeros in a whole number, it depends on whether the number has been rounded or is exact.

The rules for rounding remain the same: that is, round up if 5 or more, do not round up if 4 or less.

ONLINE

Use the 'Significant number calculator' to adapt any number to the relevant number of significant figures: www.brightredbooks.net/N5Maths (try changing the population of the United Kingdom from 2011 to 3 sig. figs).

EXAMPLE:

Round
(a) 33 528·746 to 2 sig. figs;
(b) 0·002 797 to 3 sig. figs.

SOLUTION:

(a) 33 528·746 (which has 8 sig. figs) lies between 33 000 and 34 000.
    As the third significant figure is 5, we round up to 34 000.
    NOTE: Do not write down 34 000·0, which would have 6 sig. figs.
(b) 0·002 797 (which has 4 sig. figs) lies between 0·002 79 and 0·002 80.
    As the fourth significant figure is 7, we round up to 0·002 80.

ONLINE TEST

Take the 'Significant Figures' test at www.brightredbooks.net/N5Maths

DON'T FORGET

Zeros which are used to indicate the position of the decimal point are not significant.

EXAMPLE:

A cylinder has height 25 centimetres and radius 6 centimetres.
Calculate its volume. Give your answer correct to two sig. figs.
(Volume of a cylinder = $\pi r^2 h$)

SOLUTION:

$V = \pi r^2 h = \pi \times 6 \times 6 \times 25 = 2827·433\,388$
Hence, volume of cylinder = 2800 cm³ correct to 2 sig. figs.

 THINGS TO DO AND THINK ABOUT

When carrying out calculations where a rounded answer is required, **do not** round at each step of the calculation. This leads to an inaccurate final answer. You should only round at the end of the calculations.

SURDS

RATIONAL, IRRATIONAL AND REAL NUMBERS

Numbers which can be made by dividing one integer by another (but not dividing by zero) are called **rational numbers**. Such numbers can be expressed as fractions: for example, $\frac{3}{5}$, 2 ($= \frac{2}{1}$), $0\cdot666\,666\,\ldots$ ($= \frac{2}{3}$), $-\frac{2}{5}$, $4\frac{1}{2}$ ($= \frac{9}{2}$), and so on. The set of all rational numbers is represented by the letter Q.

Some numbers, however, cannot be represented by a fraction. Examples of such numbers are $\sqrt{3}$, π and $\sqrt[3]{10}$. These numbers are called **irrational numbers**.

Together, all the rational and irrational numbers form the set of **real numbers**. The set of all real numbers is represented by the letter R. Real numbers can be positive, negative or zero.

WHAT IS A SURD?

A surd is an irrational number which is the root of an integer. Examples of surds are $\sqrt{2}$, $\sqrt{3}$ or $\sqrt[3]{10}$. We will only look in National 5 Mathematics at surds which are square roots. Numbers such as $\sqrt{4}$ and $\sqrt{9}$ look like surds, but they are not, as $\sqrt{4} = 2$ and $\sqrt{9} = 3$ and of course, 2 and 3 are rational numbers.

EXAMPLE:

Which of the following numbers are surds?
$\sqrt{12}, \quad \sqrt[3]{8}, \quad \sqrt{49}, \quad \sqrt{5}, \quad \sqrt{20}, \quad \sqrt{121}$

SOLUTION:

$\sqrt{12}, \quad \sqrt{5}, \quad \sqrt{20}$ are surds. The others are not surds because $\sqrt[3]{8} = 2$, $\sqrt{49} = 7$ and $\sqrt{121} = 11$.

DON'T FORGET

When you are simplifying a surd, remember to find a factor which is a perfect square (4, 9, 16, 25 ...) and always choose the *largest* factor if there is a choice.

SIMPLIFYING SURDS

You will often be required to simplify a surd *or* express a surd in its simplest form. In order to do this, you must find a factor of the integer which is a perfect square.

The first few perfect squares to consider are 4, 9, 16, 25, 36, 49, 64, 81 and 100.

If an integer has more than one factor which is a perfect square, choose the larger or largest one.

EXAMPLE:

(a) Simplify $\sqrt{20}$
(b) Express $\sqrt{32}$ in its simplest form.

SOLUTION:

(a) $\sqrt{20} = \sqrt{4 \times 5} = \sqrt{4} \times \sqrt{5} = 2\sqrt{5}$
(b) $\sqrt{32} = \sqrt{16 \times 2} = \sqrt{16} \times \sqrt{2} = 4\sqrt{2}$.

BEWARE: in part (b), it would be easy to write
$\sqrt{32} = \sqrt{4 \times 8} = \sqrt{4} \times \sqrt{8} = 2\sqrt{8}$. While this is true, $\sqrt{32}$ has not been *fully* simplified, as $2\sqrt{8} = 2 \times \sqrt{4 \times 2} = 2 \times \sqrt{4} \times \sqrt{2} = 4\sqrt{2}$.

SURDS IN CONTEXT

Surds occur frequently in many areas of mathematics. For example, suppose you were asked to solve the equation $x^2 - 7 = 0$. You would probably go on to say that $x^2 = 7$ and therefore $x = \pm\sqrt{7}$. In the solution, $\sqrt{7}$ is a surd.

In fact, any situation in mathematics involving square roots will probably involve surds. Consider the following example.

EXAMPLE:

Find the value of x.
Express your answer as a surd in its simplest form.

SOLUTION:

This problem can be solved using Pythagoras' Theorem.
$x^2 = 12^2 - 9^2 = 144 - 81 = 63$
$\Rightarrow x = \sqrt{63} = \sqrt{9 \times 7}$
$\qquad = \sqrt{9} \times \sqrt{7} = 3\sqrt{7}.$

ADDING AND SUBTRACTING SURDS

In the same way that $3x + 2x = 5x$, then $3\sqrt{7} + 2\sqrt{7} = 5\sqrt{7}$. In other words, surds can be collected in the same way as like terms. Similarly, $\sqrt{2} + \sqrt{3}$ cannot be simplified, as the surds are not like terms. Often, surds will have to be simplified first in order to find which are like terms.

EXAMPLE:

(a) Simplify $\sqrt{12} + \sqrt{75}$
(b) Simplify $6\sqrt{20} - 2\sqrt{45} + \sqrt{5}$.

SOLUTION:

(a) $\sqrt{12} + \sqrt{75} = \sqrt{4 \times 3} + \sqrt{25 \times 3} = 2\sqrt{3} + 5\sqrt{3} = 7\sqrt{3}$
(b) $6\sqrt{20} - 2\sqrt{45} + \sqrt{5} = 6\sqrt{4 \times 5} - 2\sqrt{9 \times 5} + \sqrt{5} = 6 \times 2\sqrt{5} - 2 \times 3\sqrt{5} + \sqrt{5}$
$\qquad = 12\sqrt{5} - 6\sqrt{5} + \sqrt{5}$
$\qquad = 7\sqrt{5}.$

HINT: In part (b), one surd ($\sqrt{5}$) did not have to be simplified. This gives a strong clue that $\sqrt{5}$ will occur in the surds which do have to be simplified.

ONLINE TEST

For more practice on surds, go online and test yourself at 'Surds': www.brightredbooks.net/N5Maths

DON'T FORGET

The topic of surds is an area of mathematics which is normally tested in a non-calculator environment. Therefore, when you are studying surds, *do not* use a calculator.

VIDEO LINK

Watch 'How to simplify surds (Part 1)' online at www.brightredbooks.net/N5Maths

THINGS TO DO AND THINK ABOUT

Express each of the following in its simplest form:

(a) $\sqrt{72}$ (b) $\sqrt{50} + \sqrt{8}$ (c) $\sqrt{24} + \sqrt{6} + \sqrt{54}$ (d) $2\sqrt{28} + 5\sqrt{7} - \sqrt{63}$.

MORE SURDS

MULTIPLYING SURDS

We have seen that $\sqrt{4 \times 5} = \sqrt{4} \times \sqrt{5}$. This type of process can be used to simplify expressions involving the multiplication of surds. In general, $\sqrt{a} \times \sqrt{b} = \sqrt{a \times b}$. For example, if you were asked to simplify $\sqrt{3} \times \sqrt{7}$, then the answer would equal $\sqrt{3 \times 7}$, that is $\sqrt{21}$. Sometimes the resulting surd will have to be simplified: for example, $\sqrt{2} \times \sqrt{14} = \sqrt{2 \times 14} = \sqrt{28} = \sqrt{4 \times 7} = 2\sqrt{7}$.

DON'T FORGET

If you are multiplying surds, look for the simplest method. That is, simplify first (if possible) then multiply, or multiply first then simplify (if possible). In general, if the product of the surds is a large number, you are better to simplify first.

EXAMPLE:

Simplify $\sqrt{12} \times \sqrt{8}$.

SOLUTION:

There are two possible methods of doing this. One method is to simplify the surds first, then multiply:

$\sqrt{12} \times \sqrt{8} = \sqrt{4 \times 3} \times \sqrt{4 \times 2} = 2\sqrt{3} \times 2\sqrt{2} = 2 \times 2 \times \sqrt{3} \times \sqrt{2} = 4\sqrt{6}$.

The second method is to multiply the surds first, then simplify:
$\sqrt{12} \times \sqrt{8} = \sqrt{12 \times 8} = \sqrt{96} = \sqrt{16 \times 6} = 4\sqrt{6}$.

In the second method, notice that after multiplying 12 by 8 we had to simplify $\sqrt{96}$. This is rather tricky, as it is not immediately obvious that 16 is a factor of 96. Therefore the first method would be preferred in this case. Both methods can be used, but sometimes one is simpler than the other.

VIDEO LINK

Watch 'How to simplify surds (Part 2)' at www. brightredbooks.net/N5Maths

FURTHER EXAMPLES

EXAMPLE:

(a) Simplify $\sqrt{2}(\sqrt{3} + 4\sqrt{2})$
(b) If $\sqrt{x} = 4\sqrt{3}$, find x.

SOLUTION:

(a) $\sqrt{2}(\sqrt{3} + 4\sqrt{2}) = (\sqrt{2} \times \sqrt{3}) + (\sqrt{2} \times 4\sqrt{2}) = \sqrt{6} + 4\sqrt{4} = \sqrt{6} + 8$
(b) $\sqrt{x} = 4\sqrt{3} \Rightarrow \sqrt{x} = \sqrt{16} \times \sqrt{3} = \sqrt{48} \Rightarrow x = 48$.

DIVIDING SURDS

Some division of surds is straightforward. A general rule for dividing surds is $\sqrt{a} \div \sqrt{b} = \sqrt{a \div b}$ or $\frac{\sqrt{a}}{\sqrt{b}} = \sqrt{\frac{a}{b}}$.

EXAMPLE:

Simplify $\frac{\sqrt{24}}{\sqrt{3}}$.

SOLUTION:

$\frac{\sqrt{24}}{\sqrt{3}} = \sqrt{\frac{24}{3}} = \sqrt{8} = \sqrt{4 \times 2} = 2\sqrt{2}$.

RATIONALISING THE DENOMINATOR

When dealing with fractions involving surds, it is usual for the denominator of the fraction to be expressed as a rational number. The process of doing this is known as rationalising the denominator. It relies upon the fact that when you multiply a surd by itself you get a rational number, that is, $\sqrt{2} \times \sqrt{2} = 2$, $\sqrt{3} \times \sqrt{3} = 3$ and so on. In general, $\sqrt{a} \times \sqrt{a} = a$. Therefore, when $\sqrt{a}$ appears in the denominator of a fraction, we multiply the fraction, *top and bottom*, by $\sqrt{a}$ to rationalise the denominator. Note that as $\frac{\sqrt{a}}{\sqrt{a}} = 1$, we are multiplying the fraction by 1 and not altering its value.

> **EXAMPLE:**
>
> (a) Express $\frac{4}{\sqrt{5}}$ as a fraction with a rational denominator.
>
> (b) Express $\frac{12}{\sqrt{2}}$ as a fraction with a rational denominator.
>
> **SOLUTION:**
>
> (a) $\frac{4}{\sqrt{5}} = \frac{4}{\sqrt{5}} \times \frac{\sqrt{5}}{\sqrt{5}} = \frac{4\sqrt{5}}{5}$
>
> (b) $\frac{12}{\sqrt{2}} = \frac{12}{\sqrt{2}} \times \frac{\sqrt{2}}{\sqrt{2}} = \frac{12\sqrt{2}}{2} = 6\sqrt{2}$
>
> NOTE: In both parts, we started with a surd (an irrational number) on the denominator, and, after our process, the denominator was rationalised, to 5 in part (a), and to 2 in part (b), although we were able to divide 2 into 12 to simplify the fraction further.

To end the section on surds, we should look at a more difficult example:

> **EXAMPLE:**
> Solve, for x, $\sqrt{x} + \sqrt{12} = 7\sqrt{3}$.
>
> **SOLUTION:**
>
> If $\sqrt{x} + \sqrt{12} = 7\sqrt{3}$, then $\sqrt{x} = 7\sqrt{3} - \sqrt{12}$.
> Now simplify $\sqrt{12}$.
> This leads to $\sqrt{x} = 7\sqrt{3} - \sqrt{4 \times 3} \Rightarrow \sqrt{x} = 7\sqrt{3} - 2\sqrt{3} = 5\sqrt{3}$.
> Hence $\sqrt{x} = 5\sqrt{3} \Rightarrow x = (5\sqrt{3})^2 = 5\sqrt{3} \times 5\sqrt{3} = 25 \times 3 = 75$.

DON'T FORGET

If you are asked to rationalise the denominator of a fraction where $\sqrt{a}$ appears on the denominator, multiply top and bottom by $\sqrt{a}$.

VIDEO LINK

Watch 'How to rationalise the denominator with surds' at www.brightredbooks.net/N5Maths

ONLINE TEST

For even more practice on surds, go online and test yourself at 'Surds': www.brightredbooks.net/N5Maths

SUMMARY

After studying surds, you should be able to:

- simplify a surd
- simplify expressions involving addition, subtraction, multiplication and division of surds
- rationalise the denominator.

THINGS TO DO AND THINK ABOUT

(a) Simplify $\sqrt{2} \times \sqrt{5}$

(b) Simplify $\sqrt{6} \times \sqrt{2}$

(c) Simplify $\sqrt{2}(\sqrt{5} + 3\sqrt{2})$

(d) Simplify $\frac{\sqrt{40}}{\sqrt{2}}$

(e) If $\sqrt{x} = 5\sqrt{2}$, find x

(f) Express $\frac{12}{\sqrt{3}}$ as a fraction with a rational denominator

(g) Solve, for x, $\sqrt{x} + \sqrt{12} = 5\sqrt{3}$.

INDICES

You should know that a^5 is a short way of writing $a \times a \times a \times a \times a$. In an expression such as this, a is called the *base* and 5 is called the *index*. The plural of index is *indices*. When we read a^5, we say 'a to the power 5'.

THE LAWS OF INDICES

The first law

Suppose we have to multiply $a^5 \times a^3$. This must equal $(a \times a \times a \times a \times a) \times (a \times a \times a) = a^8$.

This shows that when you *multiply* numbers raised to powers, you *add* indices.

Hence, the first law of indices is $a^p \times a^q = a^{p+q}$.

DON'T FORGET

You must know the laws of indices in order to tackle this topic successfully.

The second law

Suppose we have to divide $a^5 \div a^3$. This must equal $\frac{a \times a \times a \times a \times a}{a \times a \times a} = a^2$.

This shows that when you *divide* numbers raised to powers, you *subtract* indices.

Hence, the second law of indices is $a^p \div a^q = a^{p-q}$.

The third law

Suppose we have to simplify $(a^2)^3$. This must equal $(a \times a) \times (a \times a) \times (a \times a) = a^6$.

This shows that when you *raise* a number to a power to another power, you *multiply* indices.

Hence, the third law of indices is $(a^p)^q = a^{pq}$.

The fourth law

Suppose we have to simplify $(ab)^2$. This must equal $ab \times ab = a \times a \times b \times b = a^2 b^2$.

Hence, the fourth law of indices is $(ab)^n = a^n b^n$.

USING THE LAWS OF INDICES

EXAMPLE:

Simplify (a) $(p^3)^4 \times p^2$

SOLUTION:

(a) $(p^3)^4 \times p^2 = p^{3 \times 4} \times p^2 = p^{12} \times p^2 = p^{12+2} = p^{14}$ (Laws 3 and 1)

EXAMPLE:

Simplify (b) $\frac{b^3 \times b^9}{b^7}$

SOLUTION:

(b) $\frac{b^3 \times b^9}{b^7} = \frac{b^{3+9}}{b^7} = \frac{b^{12}}{b^7} = b^{12-7} = b^5$ (Laws 1 and 2)

EXAMPLE:

Simplify (c) $(2x^3)^2$.

SOLUTION:

(c) $(2x^3)^2 = 2^2(x^3)^2 = 4x^{3 \times 2} = 4x^6$ (Laws 4 and 3)

ZERO, NEGATIVE AND RATIONAL INDICES

A zero index

We can see that $a^n \div a^n$ is obviously 1. Using the second law of indices, $a^n \div a^n = a^{n-n} = a^0$.

It is true therefore that $a^0 = 1$. You *must* remember this fact.

You should also remember that $a^1 = a$.

Negative indices

Now suppose we have to divide $a^3 \div a^5$. This must equal $\frac{a \times a \times a}{a \times a \times a \times a \times a} = \frac{1}{a^2}$.

However, $a^3 \div a^5 = a^{3-5} = a^{-2}$, using the second law of indices. Therefore $a^{-2} = \frac{1}{a^2}$.

In general, $a^{-n} = \frac{1}{a^n}$.

Rational indices

Remember that rational numbers can be expressed as fractions, so we now consider indices which are fractions, for example $x^{\frac{1}{2}}$. If we think of $\left(x^{\frac{1}{2}}\right)^2$, we can start to understand rational indices. $\left(x^{\frac{1}{2}}\right)^2 = x^{\frac{1}{2}} \times x^{\frac{1}{2}} = x^{\frac{1}{2} + \frac{1}{2}} = x^1 = x$, using the first law of indices, hence $x^{\frac{1}{2}} = \sqrt{x}$. In a similar way, it can be shown that $x^{\frac{1}{3}} = \sqrt[3]{x}$ and $x^{\frac{2}{3}} = \left(\sqrt[3]{x}\right)^2$.

In general, $x^{\frac{m}{n}} = \left(\sqrt[n]{x}\right)^m$.

In view of the roots involved, there is a relationship between rational indices and surds.

ONLINE TEST

How much do you know about indices? Test yourself by visiting 'Indices' at www.brightredbooks.net/N5Maths

DON'T FORGET

You must remember that $a^0 = 1$ and also learn the general rules for both negative and rational indices.

USING ZERO, NEGATIVE AND RATIONAL INDICES

EXAMPLE:

(a) Simplify $d^3 (d^4 + d^{-3})$

SOLUTION:

(a) $d^3(d^4 + d^{-3}) = d^3 \times d^4 + d^3 \times d^{-3} = d^{3+4} + d^{3+(-3)} = d^7 + d^0 = d^7 + 1$.

EXAMPLE:

(b) Simplify $n^3 \times n^{-7}$, giving your answer with a positive power

SOLUTION:

(b) $n^3 \times n^{-7} = n^{3+(-7)} = n^{-4} = \frac{1}{n^4}$.

EXAMPLE:

(c) Evaluate $8^{\frac{2}{3}}$

SOLUTION:

(c) $8^{\frac{2}{3}} = \left(\sqrt[3]{8}\right)^2 = 2^2 = 4$.

NOTE: In examples like part (c), always evaluate the root part first, that is the cube root of 8 is 2, then evaluate the power, that is 2 to the power 2 (or 2 squared) is 4.

ONLINE

Practise indices further online by clicking 'Indices Revision' at www.brightredbooks.net/N5Maths

THINGS TO DO AND THINK ABOUT

(a) Simplify $z^5 \times (z^3)^4$

(b) Simplify $\frac{y^3 \times y^{10}}{y^7 \times y}$

(c) Simplify $(x^5 y)^3$

(d) Simplify $m^2(m^5 + m^{-2})$

(e) Simplify $n^{-4} \times n^{-1}$, giving your answer with a positive power

(f) Evaluate $25^{\frac{3}{2}}$.

SCIENTIFIC NOTATION

You should be familiar with the process of writing numbers in **scientific notation**. Scientific notation, also known as **standard form**, is a way of writing very large or very small numbers in a more convenient form.

As an example, we can consider the distance from Earth to Saturn. The distance is approximately 1·277 billion kilometres. In full, this number is 1 277 000 000. When written in scientific notation, the distance is $1·277 \times 10^9$ kilometres, a much more compact form.

REMINDERS

When we write a number in scientific notation (or standard form), we are expressing the number in the form $a \times 10^n$, where a is a number between 1 and 10 (sometimes written as $1 \leqslant a < 10$) and n is an integer. The integer is negative if the number is less than 1.

In the case of 1 277 000 000, to write it in scientific notation, we think of it as 1 277 000 000·0, then move the decimal point to the left in order to get a number between 1 and 10 (1·277). We count that the point moved 9 places, leading to $1·277 \times 10^9$.

For very small numbers, such as the probability of winning the lottery jackpot, which is 0·000 000 0715, we move the decimal point to the right in order to get a number between 1 and 10 (7·15). We count that the point moved 8 places, leading to $7·15 \times 10^{-8}$.

We can reverse the process if we wish to express numbers in scientific notation as normal numbers.

DON'T FORGET

When a number is written in scientific notation, it appears in the form $a \times 10^n$, where a is a number between 1 and 10 and n is an integer.

EXAMPLE:

Express (a) $5·6 \times 10^4$ (b) $2·93 \times 10^{-6}$ as normal numbers

SOLUTION:

(a) 56 000 (b) 0·000 002 93

On many occasions, you will require a calculator for work on scientific notation; however, you should also be able to carry out some calculations **without** a calculator. This could involve using the laws of indices.

EXAMPLE:

Express $(7 \times 10^4) \times (3 \times 10^5)$ in the form $a \times 10^n$, where $1 \leqslant a < 10$ and n is an integer.

SOLUTION:

$(7 \times 10^4) \times (3 \times 10^5) = 7 \times 3 \times 10^4 \times 10^5 = 21 \times 10^{4+5} = 21 \times 10^9$
$= 2·1 \times 10^1 \times 10^9$
$= 2·1 \times 10^{10}.$

SCIENTIFIC NOTATION — USING A CALCULATOR

Many very large or very small numbers have too many digits to fit into a calculator. For example, in science, a number known as the Avogadro constant is approximately $6·02 \times 10^{23}$ and would have 24 digits if written in full. To cope with this, we use the EXP key on the calculator. For the Avogadro constant, key in 6·02 EXP 23. You should see the following, depending on the calculator.

contd

$$6.02 \quad 23 \quad \text{or} \quad 6.02 \times 10^{23} \quad \text{or} \quad 6.02 \text{ E } 23$$

If you have a negative power of 10, for example 3.1×10^{-6}, key in 3·1 EXP (—) 6. On some calculators, the +/– key must be used.

Once you are familiar with the EXP key, you will be able to carry out more complicated problems on your calculator. As an example, look at the non-calculator example from the previous page.

For $(7 \times 10^4) \times (3 \times 10^5)$, you would key in 7 EXP 4 × 3 EXP 5. You should see 2·1 E 10 or equivalent, meaning 2.1×10^{10}.

To complete this section, we shall look at two difficult examples on scientific notation.

EXAMPLE:

The orbit of the Earth is circular.

The radius of the orbit is 1.5×10^8 kilometres.

Calculate the circumference of the orbit.

Give your answer in **scientific notation**.

SOLUTION:

$C = \pi d = \pi \times 2 \times 1.5 \times 10^8$.

Now key in $\pi \times 2 \times 1.5$ EXP 8. This leads to 942 477 796·1 on the screen.

Express this in scientific notation, that is, 9.4×10^8 kilometres.

NOTE: As the radius was given correct to 2 sig. figs (1.5×10^8), then the solution should also be rounded to 2 sig. figs.

EXAMPLE:

The total mass of the Earth is 5.97×10^{24} kilograms.

The mass of water on the Earth's surface is 1.35×10^{21} kilograms.

Express the mass of water on the Earth's surface as a percentage of the total mass of the Earth.

Give your answer in scientific notation.

SOLUTION:

Percentage = $\frac{1.35 \times 10^{21}}{5.97 \times 10^{24}} \times 100$.

Now key in 1·35 EXP 21 ÷ 5·97 EXP 24 × 100. This leads to 0·022 613 065 33.

Therefore the percentage is 2.26×10^{-2} in scientific notation (to 3 sig. figs).

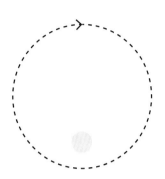

DON'T FORGET

Learn how to use the EXP key on your calculator to solve difficult problems on scientific notation.

VIDEO LINK

For more on scientific notation, watch the 'Scientific Notation Tutorial': www.brightredbooks.net/N5Maths

ONLINE TEST

Take the 'Scientific Notation' test at www.brightredbooks.net/N5Maths

 THINGS TO DO AND THINK ABOUT

(a) Express $(8 \times 10^4) \div (5 \times 10^{-9})$ in the form $a \times 10^n$, where $1 \leqslant a < 10$ and n is an integer. Do not use a calculator.

(b) Express $(6.5 \times 10^4) \times (5.2 \times 10^{-9})$ in scientific notation.

(c) Express $(2.75 \times 10^8) \div (5.36 \times 10^{-4})$ in scientific notation.

Use a calculator for both parts.

MORE INDICES

We have now looked at the laws of indices, zero, rational and negative indices and scientific notation. In this section, we will look in more detail at indices and consider some more difficult examples.

SUMMARY

To start, here is a summary of what you should know, with illustrations.

RULE	ILLUSTRATION
1 $a^p \times a^q = a^{p+q}$	$a^7 \times a^3 = a^{7+3} = a^{10}$
2 $a^p \div a^q = a^{p-q}$	$x^{12} \div x^7 = x^{12-7} = x^5$
3 $(a^p)^q = a^{pq}$	$(y^6)^2 = y^{6 \times 2} = y^{12}$
4 $(ab)^n = a^n b^n$	$(pq^2)^5 = p^5(q^2)^5 = p^5 q^{2 \times 5} = p^5 q^{10}$
5 $a^0 = 1$	$3^0 = 1$
6 $a^{-n} = \frac{1}{a^n}$	$k^{-4} = \frac{1}{k^4}$
7 $a^{\frac{m}{n}} = \left(\sqrt[n]{a}\right)^m$	$m^{\frac{3}{4}} = \left(\sqrt[4]{m}\right)^3$

When the first four laws of indices are combined with rational and negative indices, difficulties start to appear, especially as this type of question will have to be tackled **without** a calculator. Consider the example from the first section on indices, where we found that $8^{\frac{1}{3}} = 4$. This required us to calculate the cube root of 8. Of course, $\sqrt[3]{8} = 2$ because $2 \times 2 \times 2$ or $2^3 = 8$. It will help you to evaluate rational indices if you remember a few basic roots. You should already know basic square roots. Check that the first few cube roots are $\sqrt[3]{8} = 2$, $\sqrt[3]{27} = 3$, $\sqrt[3]{64} = 4$ and $\sqrt[3]{125} = 5$. You should also remember that, as $2 \times 2 \times 2 \times 2$ or $2^4 = 16$, then the fourth root of 16 is 2. This is written as $\sqrt[4]{16} = 2$.

Using these ideas, we will evaluate $16^{\frac{3}{2}}$. Now $16^{\frac{3}{2}} = \left(\sqrt[2]{16}\right)^3$. The symbol $\sqrt[2]{}$ is the same as $\sqrt{}$ and means 'square root'. Therefore we work out the square root of 16, that is 4, and then work out $4^3 = 4 \times 4 \times 4 = 64$. Hence $16^{\frac{3}{2}} = 64$. Remember *always* to work out the root first, then the power. You get the same answer doing the calculation the other way round, but it is much more difficult.

Combining points 6 and 7 from the summary also leads to difficult examples.

ONLINE

For more information on indices, visit 'Maths Revision: Indices' at www.brightredbooks.net/N5Maths

EXAMPLE:

Evaluate $16^{-\frac{3}{4}}$.

SOLUTION:

$16^{-\frac{3}{4}} = \frac{1}{16^{\frac{3}{4}}} = \frac{1}{\left(\sqrt[4]{16}\right)^3} = \frac{1}{2^3} = \frac{1}{8}$.

In this case, you should deal with the negative index first, using point 6 from the summary, then evaluate the denominator of the fraction using point 7 from the summary.

EXAMPLES WITH CONSTANTS

Look at the following example. Simplify $\frac{6x^5 \times 4x^3}{3x^6}$.

The constants (6, 4 and 3) can be dealt with separately from the terms with indices.

$$\frac{6x^5 \times 4x^3}{3x^6} = \frac{6 \times 4 \times x^5 \times x^3}{3x^6} = \frac{6 \times 4 \times x^{5+3}}{3x^6} = \frac{24x^8}{3x^6} = \frac{24}{3} \times \frac{x^8}{x^6} = 8 \times x^{8-6} = 8x^2.$$

Normally, some of these steps can be omitted, but they are included here for completeness.

RATIONAL INDICES

The presence of fractions means that examples on the laws of indices involving fractions can be particularly difficult, especially if you are unsure how to add, subtract and multiply fractions. While we will look at fractions in great detail later in the guide, it is hoped you will be able to cope at present. Study the examples below, checking the fraction calculations very carefully.

EXAMPLE:

Simplify (a) $p^{\frac{5}{2}} \times p^{\frac{1}{2}}$

SOLUTION:

(a) $p^{\frac{5}{2}} \times p^{\frac{1}{2}} = p^{\frac{5}{2}+\frac{1}{2}} = p^{\frac{6}{2}} = p^3$.

EXAMPLE:

Simplify (b) $f^{\frac{2}{3}} \div f^{-\frac{4}{3}}$

SOLUTION:

(b) $f^{\frac{2}{3}} \div f^{-\frac{4}{3}} = f^{\frac{2}{3}-\left(-\frac{4}{3}\right)} = f^{\frac{2}{3}+\frac{4}{3}} = f^{\frac{6}{3}} = f^2$.

EXAMPLE:

Simplify (c) $(m^{\frac{3}{2}})^4$

SOLUTION:

(c) $(m^{\frac{3}{2}})^4 = m^{\frac{3}{2} \times 4} = m^{\frac{12}{2}} = m^6$.

EXAMPLE:

Multiply out the brackets and simplify.
$c^{\frac{2}{3}}\left(c^{\frac{4}{3}} - c^{-\frac{2}{3}}\right)$

SOLUTION:

$c^{\frac{2}{3}}\left(c^{\frac{4}{3}} - c^{-\frac{2}{3}}\right) = c^{\frac{2}{3}} \times c^{\frac{4}{3}} - c^{\frac{2}{3}} \times c^{-\frac{2}{3}} = c^{\frac{2}{3}+\frac{4}{3}} - c^{\frac{2}{3}+\left(-\frac{2}{3}\right)} = c^{\frac{6}{3}} - c^0 = c^2 - 1.$

There is a lot to take in as you study this example, so take your time and go slowly through the working a step at a time.

 DON'T FORGET

Remember how to add and subtract fractions with a common denominator, for example $\frac{5}{2} + \frac{1}{2} = \frac{6}{2}$ and $\frac{5}{2} - \frac{1}{2} = \frac{4}{2}$, and also remember how to multiply a fraction by a whole number, for example $4 \times \frac{3}{2} = \frac{12}{2}$. Remember too to check if the resulting fraction can be simplified.

 ONLINE TEST

To take the 'More Indices' test, visit www.brightredbooks.net/N5Maths

 THINGS TO DO AND THINK ABOUT

(a) Evaluate $9^{-\frac{1}{2}}$

(b) Simplify $p^{\frac{7}{2}} \times p^{\frac{3}{2}}$

(c) Simplify $f^{\frac{1}{3}} \div f^{-\frac{8}{3}}$

(d) Simplify $(h^{\frac{5}{3}})^6$

(e) Simplify $\frac{x^{\frac{2}{3}} \times x^{\frac{5}{3}}}{x^{\frac{4}{3}}}$

(f) Multiply out the brackets and simplify

$c^{\frac{5}{3}}\left(c^{\frac{4}{3}} - c^{-\frac{5}{3}}\right)$.

MULTIPLYING OUT BRACKETS

In this section, we investigate multiplying out brackets. We have already seen some examples of this in the sections on surds and indices when we encountered expressions such as $\sqrt{2}(\sqrt{3} + 4\sqrt{2})$ and $d^3(d^4 + d^{-3})$.

THE DISTRIBUTIVE LAW

In order to simplify such expressions, we used the basic rule

$$a(b + c) = ab + ac$$

which is known as the distributive law.

You should already have used this law to simplify expressions such as $7(x + 4)$. You should remember that $7(x + 4) = 7 \times x + 7 \times 4 = 7x + 28$. Can you simplify $3x(2x - 5)$? It is slightly harder. Check that $3x(2x - 5) = 3x \times 2x - 3x \times 5 = 6x^2 - 15x$. Now look at the following examples.

EXAMPLE:

Multiply out the brackets and collect like terms:

(a) $3(x + 5) + 4(2x - 3)$ (b) $5(2y - 1) - 3(y - 4)$.

SOLUTION:

(a) $3(x + 5) + 4(2x - 3) = 3x + 15 + 8x - 12 = 11x + 3$

(b) $5(2y - 1) - 3(y - 4) = 10y - 5 - 3y + 12 = 7y + 7$.

Check the working carefully and note particularly in part (b) that $-3(y - 4) = -3y + 12$ because $-3 \times -4 = 12$. This reminds you to be extra careful when multiplying out a bracket which has a negative number before it, as the sign inside the bracket changes.

Many algebraic problems can be solved by multiplying out brackets.

EXAMPLE:

The rectangle and square shown below have the same perimeter.

$(x + 4)$ cm   Length

$(3x + 2)$ cm

Prove that the length of a side of the square is $(2x + 3)$ centimetres.

SOLUTION:

Perimeter of the square = 4 × Length

Perimeter of the rectangle = $2(x + 4) + 2(3x + 2) = 2x + 8 + 6x + 4 = 8x + 12$

As these are equal, $8x + 12 = 4 \times$ Length

Hence Length = $\frac{1}{4}(8x + 12)$

So, the length of a side of the square = $(2x + 3)$ cm.

NOTE: This type of question is known as a proof, where you are given the solution and asked to prove that the given answer is correct. Because you are given the answer, you must be very accurate and detailed in your working. Other strategies could be used in this proof.

MULTIPLYING PAIRS OF BRACKETS

The next part of the course involves multiplying pairs of brackets together such as $(x + 2)(x + 5)$ and is vital to your future success in algebra. We shall consider the above example by concentrating on the area of a rectangle $(x + 2)$ units long by $(x + 5)$ units broad:

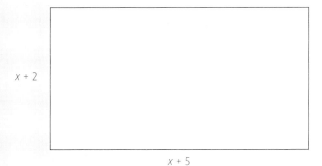

You can see that the area of the rectangle, which is $(x + 2)(x + 5)$, has been divided into four separate parts and that $(x + 2)(x + 5) = x^2 + 5x + 2x + 10 = x^2 + 7x + 10$.

There is, however, a neat way of multiplying the pair of brackets together without using a diagram.

We use the method $(x + 2)(x + 5) = x(x + 5) + 2(x + 5) = x^2 + 5x + 2x + 10 = x^2 + 7x + 10$.

Basically, every term in the first bracket gets multiplied by every term in the second bracket. So, as there are 2 terms in each bracket, then there are $2 \times 2 = 4$ terms when multiplied out before we collect like terms.

Study this systematic approach, then check the following examples:

EXAMPLE:

Simplify (a) $(x + 3)(x - 5)$
       (b) $(3x - 2)(x + 4)$
       (c) $(2a - 7)(a - 3)$.

SOLUTION:

(a) $(x + 3)(x - 5) = x(x - 5) + 3(x - 5) = x^2 - 5x + 3x - 15 = x^2 - 2x - 15$

(b) $(3x - 2)(x + 4) = 3x(x + 4) - 2(x + 4) = 3x^2 + 12x - 2x - 8 = 3x^2 + 10x - 8$

(c) $(2a - 7)(a - 3) = 2a(a - 3) - 7(a - 3) = 2a^2 - 6a - 7a + 21 = 2a^2 - 13a + 21$.

Check all the signs carefully and make sure you follow all the working. You should be aware that there are different approaches to teaching this process, some involving a diagram, some involving a mnemonic such as FOIL. As long as you can consistently produce the correct answer, the method you use will not matter.

VIDEO LINK

Watch the 'Quick tutorial on multiplying out pairs of brackets' at www.brightredbooks.net/N5Maths

DON'T FORGET

Multiply every term in the first bracket by every term in the second bracket.

THINGS TO DO AND THINK ABOUT

Multiply out the brackets and collect like terms.

(a) $(y + 3)(y + 1)$ (b) $(p + 1)(3p - 4)$

(c) $(4x - 5)(2x + 3)$ (d) $(5b - 2)(b - 4)$.

MORE BRACKETS

SQUARING BRACKETS

The process of multiplying out brackets can be described in different ways. For example, you could be asked to 'multiply out the brackets and collect like terms' or, more simply, 'simplify' or 'expand the brackets'. We are now going to look at squaring a bracket, for example $(a + b)^2$ or $(a - b)^2$. If you remember that to square something, you multiply it by itself and follow the rules from the previous section, you will arrive at a solution.

$(a + b)^2 = (a + b)(a + b) = a(a + b) + b(a + b) = a^2 + ab + ab + b^2 = a^2 + 2ab + b^2$

$(a - b)^2 = (a - b)(a - b) = a(a - b) - b(a - b) = a^2 - ab - ab + b^2 = a^2 - 2ab + b^2$.

Study the pattern for each example, checking the signs carefully. We have two important results for squaring a bracket, one with a plus sign and one with a minus sign. These results,

$$(a + b)^2 = a^2 + 2ab + b^2 \quad \textbf{and} \quad (a - b)^2 = a^2 - 2ab + b^2,$$

provide a template for squaring all brackets containing two terms. You will notice that when you square a bracket containing two terms, there are always **three** terms in the solution.

> **EXAMPLE:**
> Simplify   (a) $(x + 5)^2$ (b) $(2y - 7)^2$.
>
> **SOLUTION:**
> (a) $(x + 5)^2 = (x + 5)(x + 5) = x(x + 5) + 5(x + 5) = x^2 + 5x + 5x + 25$
> $= x^2 + 10x + 25$
>
> (b) $(2y - 7)^2 = (2y - 7)(2y - 7) = 2y(2y - 7) - 7(2y - 7) = 4y^2 - 14y - 14y + 49$
> $= 4y^2 - 28y + 49$.

Once you become very accurate and confident about multiplying out brackets, it is possible to use the two templates from the start of the section to arrive at the solution without so many stages.

For example, using $(a + b)^2 = a^2 + 2ab + b^2$ as a template, then

$$(x + 5)^2 = x^2 + (2 \times x \times 5) + 5^2 = x^2 + 10x + 25.$$

However, the most important thing is to get the correct solution. It is advisable to take your time and use a more methodical approach, as indicated earlier. The method shown above should only be tried if you have complete mastery of the more methodical method.

MORE DIFFICULT EXAMPLES

> **EXAMPLE:**
> Simplify   (a) $8x + (x - 6)(x + 2)$ (b) $(2x - 1)^2 - 3(x^2 - 5)$.
>
> **SOLUTION:**
> (a) $8x + (x - 6)(x + 2) = 8x + x(x + 2) - 6(x + 2)$
> $= 8x + x^2 + 2x - 6x - 12$
> $= x^2 + 4x - 12$.
>
> (b) $(2x - 1)^2 - 3(x^2 - 5) = (2x - 1)(2x - 1) - 3(x^2 - 5)$
> $= 2x(2x - 1) - 1(2x - 1) - 3(x^2 - 5)$
> $= 4x^2 - 2x - 2x + 1 - 3x^2 + 15$
> $= x^2 - 4x + 16$.

Please check both examples carefully, checking all the processes and paying particular attention to the plus and minus signs.

MORE TERMS IN A BRACKET

The final section on brackets involves the case where there are two terms in one bracket and three terms in the other bracket, that is an expression of the form $(ax + b)(cx^2 + dx + e)$.

To multiply out the brackets in this case, we follow the same principles as before by using a systematic approach and making sure that every term in the first bracket gets multiplied by every term in the second bracket. So, as there are 2 terms in the first bracket and 3 terms in the second bracket, there are $2 \times 3 = 6$ terms when multiplied out before we collect like terms.

EXAMPLE:

Simplify (a) $(x + 2)(3x^2 + 2x - 4)$ (b) $(3x - 5)(2x^2 - x + 7)$.

SOLUTION:

(a) $(x + 2)(3x^2 + 2x - 4) = x(3x^2 + 2x - 4) + 2(3x^2 + 2x - 4)$
$= 3x^3 + 2x^2 - 4x + 6x^2 + 4x - 8$
$= 3x^3 + 8x^2 - 8$.

(b) $(3x - 5)(2x^2 - x + 7) = 3x(2x^2 - x + 7) - 5(2x^2 - x + 7)$
$= 6x^3 - 3x^2 + 21x - 10x^2 + 5x - 35$
$= 6x^3 - 13x^2 + 26x - 35$.

HINT: Practise this type of example with the plus and minus signs in different positions until you feel confident.

VIDEO LINK

Watch 'Removing and squaring brackets' at www.brightredbooks.net/N5Maths

ONLINE TEST

Take the test 'More questions on multiplying out brackets' at www.brightredbooks.net/N5Maths

SUMMARY

After studying the two sections on multiplying out brackets, you should be able to:

- simplify $ax(bx + c)$
- simplify $a(bx + c) + d(ex + f)$
- simplify $(ax + b)(cx + d)$ including $(ax + b)^2$
- simplify $(ax + b)(cx^2 + dx + e)$

where a, b, c, d, e and f are integers.

THINGS TO DO AND THINK ABOUT

In each of the following, multiply out the brackets and simplify:

(a) $5x(3 - 2x)$

(b) $4(2x + 5) + 3(4x - 2)$

(c) $(3x + 5)(2x - 1)$

(d) $(x + 10)^2$

(e) $(4x - 5)^2$

(f) $(3x + 5)(3x - 5)$

(g) $(x + 5)(x^2 - 3x - 4)$

(h) $(3x - 1)(2x^2 + 5x - 2)$.

INTRODUCTION TO FACTORISATION

WHAT IS FACTORISATION?

We have looked at how to multiply out brackets. The reverse process to multiplying out brackets is called **factorisation**. Do you remember that $7(x + 4) = 7x + 28$ from an earlier section? Well, if we were asked to factorise $7x + 28$, then the solution would be $7(x + 4)$. One good thing about factorisation is that you can always check if your solution is correct by multiplying out the brackets (provided you can multiply out brackets accurately).

We shall look at three different methods of factorising expressions corresponding to different results from multiplying out brackets. The first involves a common factor.

COMMON FACTOR

In the example mentioned above, we looked at how to factorise $7x + 28$. To do this, we must find the **highest common factor** of $7x$ and 28. The highest common factor is 7. This common factor appears before the bracket in the solution. We can then see that the final solution must be $7(x + 4)$. If we were asked to factorise $10x^2 + 12x$, it is obvious that the highest common factor of 10 and 12 is 2; however, note that x is also a factor of $10x^2$ and $12x$, so the **highest** common factor is $2x$, leading to $10x^2 + 12x = 2x(5x + 6)$. Once the common factor has been found, it should be straightforward to complete the bracket.

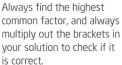

DON'T FORGET

Always find the highest common factor, and always multiply out the brackets in your solution to check if it is correct.

ONLINE TEST

Take the 'Factorisation' test at www.brightredbooks.net/N5Maths

> **EXAMPLE:**
> Factorise (a) $4x + 20$ (b) $12y - 16$ (c) $7d + 14d^2$ (d) $6x - 8y + 10$.
>
> **SOLUTION:**
> (a) $4(x + 5)$ (b) $4(3y - 4)$ (c) $7d(1 + 2d)$ (d) $2(3x - 4y + 5)$.

DIFFERENCE OF TWO SQUARES

When multiplying out pairs of brackets, we sometimes come across brackets with the same terms but opposite signs, such as $(a + b)(a - b)$. Look at what happens when we multiply out the brackets.

$$(a + b)(a - b) = a(a - b) + b(a - b) = a^2 - ab + ab - b^2 = a^2 - b^2.$$

The two middle terms cancel one another out, leaving $(a + b)(a - b) = a^2 - b^2$. Therefore when we factorise $a^2 - b^2$, we get $(a + b)(a - b)$. This type of factorisation is called a difference of two squares, as we start with the subtraction of two terms, both of which are perfect squares.

The rule $a^2 - b^2 = (a + b)(a - b)$ can be used to factorise many related expressions.

> **EXAMPLE:**
> Factorise (a) $a^2 - 25$ (b) $p^2 - 100$.
>
> **SOLUTION:**
> (a) $a^2 - 25 = a^2 - 5^2 = (a + 5)(a - 5)$
> (b) $p^2 - 100 = p^2 - 10^2 = (p + 10)(p - 10)$.
> This type of factorisation can be extended to expressions of the type $px^2 - q^2$.

contd

EXAMPLE:

Factorise   (a) $25x^2 - 36$ (b) $4x^2 - 81$.

SOLUTION:

(a) $25x^2 - 36 = (5x)^2 - 6^2 = (5x + 6)(5x - 6)$

(b) $4x^2 - 81 = (2x)^2 - 9^2 = (2x + 9)(2x - 9)$.

COMMON FACTOR WITH DIFFERENCE OF TWO SQUARES

Suppose you are asked to factorise $2x^2 - 32$. At first glance, this looks like a difference of two squares, but when we look closer we see that neither 2 nor 32 is a perfect square. However, the expression can be factorised by using a common factor first. This leads to $2x^2 - 32 = 2(x^2 - 16)$. You then find a difference of two squares in the bracket, as $16 = 4^2$, and can continue. Hence $2x^2 - 32 = 2(x^2 - 16) = 2(x^2 - 4^2) = 2(x + 4)(x - 4)$.

EXAMPLE:

Factorise   (a) $2y^2 - 18$ (b) $4x^2 - 4$.

SOLUTION:

(a) $2y^2 - 18 = 2(y^2 - 9) = 2(y^2 - 3^2) = 2(y + 3)(y - 3)$

(b) $4x^2 - 4 = 4(x^2 - 1) = 4(x^2 - 1^2) = 4(x + 1)(x - 1)$.

DON'T FORGET

You should always check if there is a common factor in an expression first.

Now we shall consider a practical example involving the above ideas. Do not use a calculator in this example.

EXAMPLE:

A washer is made from two concentric circles with outer diameter $2R$ centimetres and inner diameter $2r$ centimetres.

(a) Show that the area of the washer is $\pi(R + r)(R - r)$ square centimetres.

(b) Calculate the area of the washer when $R = 6$ and $r = 4$. Take $\pi = 3\cdot14$.

2r cm

2R cm

SOLUTION:

(a) The area of the washer = the area of the outer circle – the area of the inner circle
$$= \pi R^2 - \pi r^2 = \pi(R^2 - r^2) = \pi(R + r)(R - r).$$

(b) Area of washer $= \pi(R + r)(R - r) = 3\cdot14 \times (6 + 4) \times (6 - 4) = 3\cdot14 \times 10 \times 2 = 62\cdot8$.
Hence area is $62\cdot8$ cm².

This was clearly a difficult example; however, did you notice in part (a) that we took out a common factor first, then had a difference of two squares?

ONLINE

For more on factorisation, watch the clip 'Factorisation' at www.brightredbooks.net/N5Maths

THINGS TO DO AND THINK ABOUT

Factorise   (a) $8x - 40$ (b) $16b + 24$ (c) $4a + 2a^2$ (d) $p^2 - 9$ (e) $y^2 - 49$

(f) $4x^2 - 9$ (g) $100h^2 - 81$ (h) $3w^2 - 12$ (i) $5x^2 - 80$.

When you are asked to factorise an expression, think about whether there is a common factor first. If there is, attend to that, then check if any further factorisation can take place, for example a difference of two squares. Once you have completed the factorisation, you can check if the solution is correct by multiplying out the brackets.

FACTORISING TRINOMIALS

When we multiplied pairs of brackets in an earlier section such as $(x + 2)(x + 5)$, we ended up with three terms in the solution, that is $x^2 + 7x + 10$. Expressions like $x^2 + 7x + 10$ are called trinomials, meaning they have three terms. When we factorise a trinomial like this, the solution should have two brackets.

HOW TO FACTORISE TRINOMIALS

Unless you know how to multiply pairs of brackets together, you will be unlikely to be able to factorise trinomials. In this section, we will study trinomials of the type $ax^2 + bx + c$, where $a = 1$. Expressions of the type $ax^2 + bx + c$ are called **quadratic expressions**.

We shall start by looking at examples where the final term is positive.

Consider how to factorise $x^2 + 7x + 10$ by thinking of how we got there initially.

$$(x + 2)(x + 5) = x(x + 5) + 2(x + 5) = x^2 + 5x + 2x + 10 = x^2 + 7x + 10$$

To factorise $x^2 + 7x + 10$, we realise that there will be two brackets and that each bracket must start with x, that is $(x \quad)(x \quad)$, to get the x^2 term. We should also realise that the product of the remaining terms in the brackets must be +10. So, we think of pairs of factors of +10. The only possibilities therefore are (1 and 10) or (–1 and –10) or (2 and 5) or (–2 and –5). The only clue to which of these four possible pairs is correct is the middle term of the trinomial, $7x$. Look through the pairs of factors for a pair which adds up to 7. This leads to (2 and 5). The solution is then $(x + 2)(x + 5)$. You should check your solution by multiplying out the brackets.

If the final term in a trinomial is **positive**, then the signs in the two brackets must be the **same**. If the middle term is positive, that means two + signs in the brackets, and if the middle term is negative, that means two – signs in the brackets.

ONLINE

For further activities on factorising trinomials, visit 'Factorising Quadratic Trinomials' at www. brightredbooks.net/N5Maths

POSITIVE FINAL TERMS

We shall now go through the steps to factorise some trinomials where the final term is positive.

EXAMPLE:

Factorise   (a) $y^2 + 9y + 20$ (b) $x^2 – 6x + 8$ (c) $a^2 + 16a + 48$

SOLUTION:

(a) Step 1   Solution must start $(y \quad)(y \quad)$
    Step 2   Think of factors of 20 which add up to 9, giving 4 and 5
    Step 3   Solution is $(y + 4)(y + 5)$. [the order of the brackets is not important]

(b) Step 1   Solution must start $(x \quad)(x \quad)$
    Step 2   Think of factors of 8 which add up to 6, giving 2 and 4
    Step 3   Solution is $(x – 2)(x – 4)$.

(c) Step 1   Solution must start $(a \quad)(a \quad)$
    Step 2   Think of factors of 48 which add up to 16, giving 4 and 12
    Step 3   Solution is $(a + 4)(a + 12)$.

NEGATIVE FINAL TERMS

Consider how to factorise $x^2 - 2x - 15$ by thinking of how we got there initially.

$(x + 3)(x - 5) = x(x - 5) + 3(x - 5) = x^2 - 5x + 3x - 15 = x^2 - 2x - 15$.

To factorise $x^2 - 2x - 15$, we realise that there will be two brackets and that each bracket must start with x, that is $(x\quad)(x\quad)$ to get the x^2 term. We should also realise that the product of the remaining terms in the brackets must be –15. So, we think of pairs of factors of –15. The only possibilities therefore are (1 and –15) or (15 and –1) or (3 and –5) or (5 and –3). The only clue to which of these four possible pairs is correct is the middle term of the trinomial, –2x. Look through the pairs of factors for a pair which adds up to –2. This leads to (–5 and 3). The solution is then $(x + 3)(x - 5)$. You should check your solution by multiplying out the brackets.

If the final term in a trinomial is **negative**, then the signs in the two brackets must be **different**. That means one + sign and one – sign in the brackets. In this case, it is simpler to think of factors of the final term which **subtract** to give the middle term and then decide where the + and – signs go.

We shall now go through the steps to factorise some trinomials where the final term is negative.

> **EXAMPLE:**
> Factorise (a) $y^2 + 8y - 20$ (b) $x^2 - 5x - 6$ (c) $a^2 - 13a - 30$
>
> **SOLUTION:**
> (a) Step 1   Solution must start $(y\quad)(y\quad)$
>   Step 2   Think of factors of 20 which subtract to give 8, leading to 2 and 10
>   Step 3   Solution must continue $(y\quad 2)(y\quad 10)$
>   Step 4   Solution is $(y - 2)(y + 10)$.   [as $-2 + 10 = +8$]
> (b) Step 1   Solution must start $(x\quad)(x\quad)$
>   Step 2   Think of factors of 6 which subtract to give 5, leading to 1 and 6
>   Step 3   Solution must continue $(x\quad 1)(x\quad 6)$
>   Step 4   Solution is $(x + 1)(x - 6)$.   [as $+1 - 6 = -5$]
> (c) Step 1   Solution must start $(a\quad)(a\quad)$
>   Step 2   Think of factors of 30 which subtract to give 13, leading to 2 and 15
>   Step 3   Solution must continue $(a\quad 2)(a\quad 15)$
>   Step 4   Solution is $(a + 2)(a - 15)$.   [as $+2 - 15 = -13$]

 ONLINE TEST

To test yourself, visit 'Factorising Trinomials' at www.brightredbooks.net/ N5Maths

 THINGS TO DO AND THINK ABOUT

Factorise   (a) $x^2 + 11x + 24$ (b) $x^2 - 8x + 15$ (c) $p^2 + 13p + 30$ (d) $y^2 - 13y + 12$

(e) $x^2 - 2x - 24$ (f) $x^2 + 7x - 18$ (g) $y^2 + 6y - 16$ (h) $y^2 - 4y - 21$.

After each solution, check it is correct by multiplying out the brackets.

You should be aware that there are different approaches to teaching this process. As long as you can consistently get the correct solution, stick to the method you are familiar with. Whatever method you are using, success at factorisation can only be achieved by practice (lots of it) and perseverance. Although it is time-consuming, it is important that you check your solutions. As you practise more, you will improve and speed up. Factorisation occurs in many future areas of mathematics, and it is essential you master how to do this.

MORE FACTORISATION

TRINOMIALS WITH A NON-UNITARY x^2 COEFFICIENT

We will now look at how to factorise more complicated trinomials where there is a number other than 1 in front of the x^2 term. The word **coefficient** refers to the number which multiplies a variable in an algebraic term: for example, 3 is the coefficient in $3x$.

We did the next example earlier in the section on multiplying out brackets.

$$(2a - 7)(a - 3) = 2a(a - 3) - 7(a - 3) = 2a^2 - 6a - 7a + 21 = 2a^2 - 13a + 21.$$

Let us think about how we could set about factorising $2a^2 - 13a + 21$. The coefficient of 2 with the a^2 term complicates matters. Again we realise that there will be two brackets and that each bracket must start with $(2a \quad)(a \quad)$ to get the $2a^2$ term. We should also realise that the product of the remaining terms in the brackets must be 21. So, we think of pairs of factors of 21. The only possibilities therefore are (1 and 21) or (3 and 7). We should also remember the statement 'If the final term in a trinomial is **positive**, then the signs in the two brackets, must be the **same**. If the middle term is positive, that means two + signs in the brackets, and if the middle term is negative, that means two – signs in the brackets.' So, it appears that the only possible solutions are $(2a - 1)$ $(a - 21)$ or $(2a - 21)(a - 1)$ or $(2a - 3)(a - 7)$ or $(2a - 7)(a - 3)$, as all four lead to $2a^2$ and +21. Only one of them, however, leads to the correct middle term, $-13a$.

outer terms

$(a + b)(c + d)$

inner terms

You should now check your solution by looking at the brackets. You know that all four possible correct solutions lead to $2a^2$ and +21, so you can concentrate on the middle two terms. This can be done by multiplying the pair of outer terms and the pair of inner terms (see diagram) from the brackets and combining them, looking for $-13a$.

Working could be set out as follows:

	Outer	Inner	Middle term
$(2a - 1)(a - 21)$	$2a \times -21 = -42a$	$-1 \times a = -a$	$-42a - a = -43a$
$(2a - 21)(a - 1)$	$2a \times -1 = -2a$	$-21 \times a = -21a$	$-2a - 21a = -23a$
$(2a - 3)(a - 7)$	$2a \times -7 = -14a$	$-3 \times a = -3a$	$-14a - 3a = -17a$
$(2a - 7)(a - 3)$	$2a \times -3 = -6a$	$-7 \times a = -7a$	$-6a - 7a = -13a$

You will see that the fourth option is correct, so the solution is $(2a - 7)(a - 3)$.

You may think this is time-consuming and will take quite a long time – and you will be right. However, as you practise more, you will see quicker ways of setting out the working, and you will be able to spot the more likely options much faster. For example, it is noticeable that the combinations using (1 and 21) are further away from the desired answer, so starting with (3 and 7) would have been better.

DON'T FORGET

With regular practice on plenty of examples, you will become expert at factorising trinomials fairly quickly. Remember to check your solution by multiplying out the brackets.

ONLINE

For even more extended activities on factorisation, visit 'Factorising Harder Quadratic Trinomials' at www.brightredbooks.net/N5Maths

EXAMPLE:

Factorise $3x^2 + 14x - 5$

SOLUTION:

Start with $(3x \quad)(x \quad)$, then check factors of 5 are 1 and 5. As the final term is negative, the signs in the two brackets must be different. The four possible solutions are $(3x - 1)(x + 5)$ or $(3x + 1)(x - 5)$ or $(3x + 5)(x - 1)$ or $(3x - 5)(x + 1)$.

	Outer	Inner	Middle term
$(3x - 1)(x + 5)$	$3x \times 5 = 15x$	$-1 \times x = -x$	$15x - x = 14x$

Right away, we find the combination leading to $+14x$ in the middle, so the correct solution is $(3x - 1)(x + 5)$. It can then be checked by multiplying out the brackets.

 contd

EXAMPLE:

Factorise $2x^2 - 7x - 15$

SOLUTION:

Start with $(2x)(x)$, then check factors of 15 are (1 and 15) or (3 and 5). As the final term is negative, the signs in the two brackets must be different. Try (3 and 5) first, as (1 and 15) may be further away from the desired answer. The four possible solutions are $(2x - 3)(x + 5)$ or $(2x + 3)(x - 5)$ or $(2x + 5)(x - 3)$ or $(2x - 5)(x + 3)$.

	Outer	Inner	Middle term
$(2x - 3)(x + 5)$	$2x \times 5 = 10x$	$-3 \times x = -3x$	$10x - 3x = 7x$
$(2x + 3)(x - 5)$	$2x \times -5 = -10x$	$3 \times x = 3x$	$-10x + 3x = -7x$

We quickly find the combination leading to $-7x$ in the middle, so the correct solution is $(2x + 3)(x - 5)$. It can then be checked by multiplying out the brackets.

Of course, if you were asked to factorise an expression such as $6x^2 + 11x - 10$, there are many possibilities of factors and this could be very time-consuming; however, you will probably be given less demanding examples to do, at least until you are more confident. The solution is given later if you feel confident enough to try.

ONLINE TEST

To test yourself, visit 'Factorising Trinomials' at www.brightredbooks.net/ N5Maths

USING A COMMON FACTOR

If the numbers are large or seem more complicated than usual, check whether there is a common factor first.

EXAMPLE:

Factorise $5x^2 - 40x + 80$

SOLUTION:

$5x^2 - 40x + 80 = 5(x^2 - 8x + 16) = 5(x - 4)(x - 4)$.

THINGS TO DO AND THINK ABOUT

Factorise   (a) $2x^2 + 11x + 5$ (b) $3x^2 - 11x - 4$ (c) $2y^2 - y - 1$

(d) $5y^2 - 16y + 3$ (e) $2x^2 + 3x - 9$ (f) $2a^2 - 9a + 10$.

The ability to factorise trinomials is vital for later work on quadratic equations. It is therefore extremely important that, firstly, you can both multiply out pairs of brackets and factorise trinomials accurately. To be able to do this, you must practise regularly and take your time.

Many examples on factorisation require patience, as the correct solution may not appear until you have tried and rejected other possibilities. So, learn to persevere at this, and the rewards will follow.

Try these examples on factorisation.

Factorise   (a) $10x + 50$ (b) $14b + 12b^2$ (c) $p^2 - 121$ (d) $25h^2 - 9$

(e) $4x^2 - 36$ (f) $x^2 + 12x + 20$ (g) $x^2 - 12x + 27$ (h) $y^2 - 14y - 15$

(i) $x^2 + 2x - 24$ (j) $2x^2 + 9x + 7$ (k) $3x^2 + 2x - 1$ (l) $3x^2 - 10x - 8$.

If you tried the difficult example from above, that is factorise $6x^2 + 11x - 10$, the solution is $(3x - 2)(2x + 5)$. Well done if you got the correct answer!

COMPLETING THE SQUARE

Earlier, we looked at examples of squaring brackets and saw that

$$(a + b)^2 = a^2 + 2ab + b^2.$$

This could be used as a template for squaring all brackets by thinking of the three terms in the solution as the square of the first term, double the product of both terms, and finally the square of the second term. Hence

$$(x + 5)^2 = x^2 + (2 \times x \times 5) + 5^2 = x^2 + 10x + 25.$$

HOW TO COMPLETE THE SQUARE

In this section, we look at how to complete the square. For example, if we were asked to add a term to $x^2 + 10x$ to make it a perfect square, what would it be? We can see that we would add 25 because $x^2 + 10x + 25 = (x + 5)^2$, which is a perfect square. Note that we can arrive at 25 by dividing 10 by 2 then squaring the result. In general, using our template, if we need to add on to $x^2 + ax$ to make a perfect square, we should divide a by 2 then square the result.

EXAMPLE:

What would you add to $x^2 + 16x$ to make a perfect square?

SOLUTION:

$16 \div 2 = 8$ and $8^2 = 64$, so add 64, as $x^2 + 16x + 64 = (x + 8)^2$.

In practice, examples would probably be given in a different form, avoiding the expression 'completing the square'. If you see a question asking you to express something in the form $(x + p)^2 + q$, then you should start thinking about completing the square.

EXAMPLE:

Express $x^2 + 16x$ in the form $(x + p)^2 + q$.

SOLUTION:

$x^2 + 16x = x^2 + 16x + 64 - 64 = (x + 8)^2 - 64$.

Note that, after adding 64 to 'complete the square', we then had to subtract 64 to maintain equality before factorising the perfect square.

EXAMPLE:

Express $x^2 + 6x + 5$ in the form $(x + p)^2 + q$.

SOLUTION:

$x^2 + 6x + 5 = x^2 + 6x + 9 - 9 + 5 = (x + 3)^2 - 9 + 5 = (x + 3)^2 - 4$.

Check through the working carefully. We divide 6 by 2 to get 3, then square 3 to get 9. After adding 9 on to $x^2 + 6x$, we have to subtract 9 to maintain equality. You can check that the solution is correct by simplifying it, that is $(x + 3)^2 - 4 = x^2 + 6x + 9 - 4 = x^2 + 6x + 5$. It is always worth checking the solution in this type of question.

ONLINE TEST

Test yourself on 'Completing Squares' by visiting www. brightredbooks.net/N5Maths

EXAMPLE:

Express $x^2 - 4x - 11$ in the form $(x + p)^2 + q$.

SOLUTION:

$x^2 - 4x - 11 = x^2 - 4x + 4 - 4 - 11 = (x - 2)^2 - 4 - 11 = (x - 2)^2 - 15$.

MAXIMUM AND MINIMUM VALUES

We can use the process of completing the square to find the maximum and minimum values of expressions of the form $ax^2 + bx + c$. Such expressions will be looked at in greater detail later in the course. For example, suppose we were asked to find the minimum value of the expression $x^2 + 6x + 5$. In other words, what number in place of x would make $x^2 + 6x + 5$ as small as possible? Well, by completing the square earlier, we found that

$$x^2 + 6x + 5 = (x + 3)^2 - 4.$$

The variable part of $(x + 3)^2 - 4$ is $(x + 3)^2$. When we square something, the minimum value it can possibly be is zero, so the minimum value of $(x + 3)^2$ is zero, and it occurs when $x = -3$. Hence the minimum value of $(x + 3)^2 - 4$ is $0 - 4 = -4$. To sum up, the minimum value of $(x + 3)^2 - 4$ is -4, and it occurs when $x = -3$.

EXAMPLE:

By expressing in the form $(x + p)^2 + q$, find the minimum value of $x^2 - 12x + 25$.

SOLUTION:

$$x^2 - 12x + 25 = x^2 - 12x + 36 - 36 + 25 = (x - 6)^2 - 36 + 25 = (x - 6)^2 - 11.$$
The minimum value, which occurs when $x = 6$, is -11.

Later on in the course, we shall draw the graphs of quadratic functions of the type

$$y = x^2 - 12x + 25.$$

The graph of this function, which is known as a parabola, is shown below.

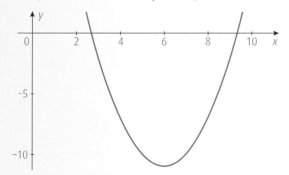

DON'T FORGET

If you are asked to express something in the form $(x + p)^2 + q$, you must use the method of completing the square. Once you have arrived at the solution, remember to check it.

We showed that the minimum value of the expression $x^2 - 12x + 25$ was -11, and it occurred when $x = 6$. This relates to the graph of $y = x^2 - 12x + 25$, as the coordinates of the minimum turning point of the graph are $(6, -11)$.

The method of completing the square can also be used to solve equations of the type $x^2 - 12x + 25 = 0$. Equations of this type are known as quadratic equations. We shall have a detailed look at quadratic functions and equations later in the course, and you will see the link between completing the square and the graphs of such functions.

VIDEO LINK

For a quick tutorial, watch 'Completing the square' at www.brightredbooks.net/N5Maths

THINGS TO DO AND THINK ABOUT

1. Express each of the following in the form $(x + p)^2 + q$.
    (a) $x^2 + 14x + 10$ (b) $x^2 - 12x + 30$ (c) $x^2 + 2x + 3$ (d) $x^2 - 8x + 12$.

Remember to check the solution each time.

2. By expressing in the form $(x + p)^2 + q$, find the minimum value of the following:
    (a) $x^2 + 6x + 3$ (b) $x^2 - 10x + 14$ (c) $x^2 + 18x + 50$ (d) $x^2 - 2x - 7$.
    (e) What are the coordinates of the minimum turning point of the parabola with equation $y = x^2 + 8x + 10$?

ALGEBRAIC FRACTIONS

SIMPLIFYING FRACTIONS

You should know how to simplify fractions such as $\frac{10}{35}$. By expressing $\frac{10}{35}$ as $\frac{5 \times 2}{5 \times 7}$ (in other words finding the factors of 10 and 35), we can simplify $\frac{10}{35}$ to $\frac{2}{7}$ by dividing the numerator and denominator of $\frac{10}{35}$ by 5 (the highest common factor of 10 and 35). We say that we cancel the 5s. This process is also known as expressing a fraction in its simplest form.

In a similar way, we can simplify algebraic fractions by cancelling factors common to the numerator and denominator of a fraction. We must use the highest common factor when doing this. For example, $\frac{pq}{pr}$ would simplify to $\frac{q}{r}$ by cancelling the p terms. Note that fractions such as $\frac{p-q}{p+r}$ cannot be simplified, due to the + and – signs. We can, however, cancel brackets common to the numerator and denominator. For example, $\frac{4(a+b)}{(a+b)(a-b)}$ would simplify to $\frac{4}{(a-b)}$ by cancelling the $(a+b)$ brackets.

> ### EXAMPLE:
>
> Express $\frac{a^2b}{ab^3}$ as a fraction in its simplest form.
>
> ### SOLUTION:
>
> $\frac{a^2b}{ab^3} = \frac{a \times a \times b}{a \times b \times b \times b} = \frac{a}{b^2}$.

In some examples, it might not be obvious that the fraction can be simplified, for example $\frac{6a+9}{12}$. However, this can be simplified if we factorise the numerator by taking out a common factor. Hence $\frac{6a+9}{12} = \frac{3(2a+3)}{12} = \frac{2a+3}{4}$.

You may have to use the skills you learned earlier to factorise a trinomial.

DON'T FORGET

To simplify an algebraic fraction, cancel factors common to the numerator and denominator. You may have to factorise the numerator and denominator first.

> ### EXAMPLE:
>
> Express $\frac{4x-8}{x^2+x-6}$ as a fraction in its simplest form.
>
> ### SOLUTION:
>
> $\frac{4x-8}{x^2+x-6} = \frac{4(x-2)}{(x-2)(x+3)} = \frac{4}{(x+3)}$.

In the next section, we shall consider the four operations with algebraic fractions – addition, subtraction, multiplication and division.

ONLINE

For a quick tutorial, visit 'Simplifying fractions' at www.brightredbooks.net/N5Maths

ADDITION

You should know how to add simple fractions without using a calculator. This is done by finding the LCM (Least Common Multiple) of the denominators of the fractions and using this as a common denominator.

> ### EXAMPLE:
>
> Calculate $\frac{3}{4} + \frac{1}{5}$
>
> ### SOLUTION:
>
> $\frac{3}{4} + \frac{1}{5} = \frac{3 \times 5}{20} + \frac{1 \times 4}{20} = \frac{15}{20} + \frac{4}{20} = \frac{19}{20}$.
> Note that we find the LCM (Least Common Multiple) of 4 and 5, which is 20 (4 × 5), and then form two equivalent fractions with denominator 20. The same method is used to add algebraic fractions.

contd

EXAMPLE:

Express $\frac{a}{b} + \frac{c}{d}$, $b \neq 0$, $d \neq 0$, as a single fraction in its simplest form.

SOLUTION:

The LCM of b and d is bd.

Hence $\frac{a}{b} + \frac{c}{d} = \frac{ad}{bd} + \frac{bc}{bd} = \frac{ad + bc}{bd}$.

Note that the two statements after the calculation, that is $b \neq 0$, $d \neq 0$, are simply there for mathematical accuracy because division by zero is impossible. They should not be considered when you are finding the solution.

EXAMPLE:

Express $\frac{2}{b} + \frac{1}{b^2}$, $b \neq 0$, as a single fraction in its simplest form.

SOLUTION:

The LCM of b and b^2 is b^2 as b is a factor of b^2.

Hence $\frac{2}{b} + \frac{1}{b^2} = \frac{2b}{b^2} + \frac{1}{b^2} = \frac{2b + 1}{b^2}$.

EXAMPLE:

Express $\frac{5}{x-3} + \frac{2}{x+2}$, $x \neq 3$, $x \neq -2$, as a single fraction in its simplest form.

SOLUTION:

The LCM of $x - 3$ and $x + 2$ is $(x - 3)(x + 2)$.

Hence $\frac{5}{x-3} + \frac{2}{x+2} = \frac{5(x+2)}{(x-3)(x+2)} + \frac{2(x-3)}{(x-3)(x+2)} = \frac{5x + 10 + 2x - 6}{(x-3)(x+2)}$
$$= \frac{7x + 4}{(x-3)(x+2)}.$$

AN EXAMPLE IN CONTEXT

EXAMPLE:

The net resistance from two resistors in series can be calculated using the formula $\frac{1}{R} = \frac{1}{R_1} + \frac{1}{R_2}$.

(a) Prove that $R = \frac{R_1 R_2}{R_1 + R_2}$

(b) Calculate R when $R_1 = 6$ and $R_2 = 4$.

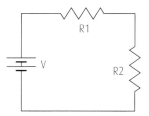

SOLUTION:

(a) $\frac{1}{R} = \frac{1}{R_1} + \frac{1}{R_2} \Rightarrow \frac{1}{R} = \frac{R_2}{R_1 R_2} + \frac{R_1}{R_1 R_2} = \frac{R_1 + R_2}{R_1 R_2} \Rightarrow R = \frac{R_1 R_2}{R_1 + R_2}$

(b) $R = \frac{R_1 R_2}{R_1 + R_2} = \frac{6 \times 4}{6 + 4} = \frac{24}{10} = 2 \cdot 4$.

ONLINE TEST

Test yourself on 'Algebraic fractions' at www.brightredbooks.net/N5Maths

THINGS TO DO AND THINK ABOUT

1. Express the following fractions in their simplest form:

(a) $\frac{a^2 b}{ab^3}$ (b) $\frac{10x^2}{5x^4}$ (c) $\frac{4y - 2}{2}$ (d) $\frac{x^2 + 7x + 12}{2x + 6}$ (e) $\frac{x^2 - 8x + 15}{x^2 + x - 12}$.

2. Express the following as single fractions in their simplest form:

(a) $\frac{4}{x} + \frac{3}{x+5}$, $x \neq 0$, $x \neq -5$ (b) $\frac{2}{x-1} + \frac{6}{x+3}$, $x \neq 1$, $x \neq -3$.

MORE FRACTIONS

SUBTRACTION

The method of subtracting fractions is the same as addition in that it is necessary to find the LCM of the denominators of the fractions and use this as a common denominator.

EXAMPLE:

Calculate $\frac{3}{4} - \frac{1}{5}$.

SOLUTION:

$$\frac{3}{4} - \frac{1}{5} = \frac{3 \times 5}{20} - \frac{1 \times 4}{20} = \frac{15}{20} - \frac{4}{20} = \frac{11}{20}.$$

Now apply this method to algebraic fractions in the same way as addition.

EXAMPLE:

Express $\frac{7}{x+4} - \frac{3}{x-1}$, $x \neq -4$, $x \neq 1$, as a single fraction in its simplest form.

SOLUTION:

The LCM of $x + 4$ and $x - 1$ is $(x + 4)(x - 1)$.

Hence $\frac{7}{x+4} - \frac{3}{x-1} = \frac{7(x-1)}{(x+4)(x-1)} - \frac{3(x+4)}{(x+4)(x-1)} = \frac{7x - 7 - 3x - 12}{(x+4)(x-1)}$

$$= \frac{4x - 19}{(x+4)(x-1)}.$$

Did you spot that $-3(x + 4) = -3x - 12$? Always be careful if you are multiplying out a bracket when subtracting algebraic fractions. If there is a bracket in the second fraction, the sign inside the bracket will change due to the minus sign before the bracket.

MULTIPLICATION

You should know how to multiply simple fractions without using a calculator.

EXAMPLE:

Calculate $\frac{2}{3} \times \frac{5}{8}$.

SOLUTION:

$$\frac{2}{3} \times \frac{5}{8} = \frac{2 \times 5}{3 \times 8} = \frac{10}{24} = \frac{2 \times 5}{2 \times 12} = \frac{5}{12}.$$

We multiply the numerators together and the denominators together, then simplify the fraction, if possible, by cancelling. You could cancel first, then multiply instead. Again, the same approach is used with algebraic fractions.

EXAMPLE:

Express $\frac{2}{x^2} \times \frac{5x}{8}$ as single fractions in their simplest form.

SOLUTION:

$$\frac{2}{x^2} \times \frac{5x}{8} = \frac{2 \times 5x}{x^2 \times 8} = \frac{2 \times 5 \times x}{x \times x \times 8} = \frac{5}{4x}.$$

DIVISION

You should know how to divide simple fractions without using a calculator. To divide two fractions, leave the first fraction as it is, change the division sign to a multiplication sign, invert the second fraction, that is turn it upside down, then do in the same way as a multiplication of fractions.

> **EXAMPLE:**
>
> Calculate $\frac{2}{9} \div \frac{1}{3}$.
>
> **SOLUTION:**
>
> $\frac{2}{9} \div \frac{1}{3} = \frac{2}{9} \times \frac{3}{1} = \frac{2 \times 3}{9 \times 1} = \frac{6}{9} = \frac{2 \times 3}{3 \times 3} = \frac{2}{3}$.

Again, the same approach is used with algebraic fractions.

> **EXAMPLE:**
>
> Calculate $\frac{3a}{4} \div \frac{a^2}{2b}$.
>
> **SOLUTION:**
>
> $\frac{3a}{4} \div \frac{a^2}{2b} = \frac{3a}{4} \times \frac{2b}{a^2} = \frac{3a \times 2b}{4 \times a^2} = \frac{3 \times a \times 2 \times b}{4 \times a \times a} = \frac{3b}{2a}$.

DON'T FORGET

When asked to divide fractions, change to a multiplication and invert the second fraction.

VIDEO LINK

Watch the clip 'Dividing Fractions' for more at www.brightredbooks.net/N5Maths

SUMMARY

After studying the two sections on algebraic fractions, you should be able to:

- simplify algebraic fractions
- add, subtract, multiply and divide simple fractions without a calculator
- add, subtract, multiply and divide algebraic fractions.

Check carefully over the four operations, as it is not unusual for students to get the techniques for addition and subtraction mixed up with the techniques for multiplication and division. The following section will give you the opportunity to practise some mixed examples.

ONLINE TEST

Test yourself on 'Algebraic fractions' at www.brightredbooks.net/N5Maths

THINGS TO DO AND THINK ABOUT

1. Express the following as single fractions in their simplest form.

   (a) $\frac{6}{x} - \frac{1}{x-4}$, $x \neq 0$, $x \neq 4$

   (b) $\frac{2}{x+6} - \frac{3}{x-1}$, $x \neq -6$, $x \neq 1$

   (c) $\frac{a}{b} \times \frac{c}{a}$ (d) $\frac{2a}{3b} \times \frac{b}{4}$ (e) $\frac{x^3}{y} \times \frac{y^2}{x^5}$ (f) $\frac{4x}{5y} \times \frac{10}{x^2}$

   (g) $\frac{p}{q} \div \frac{r}{s}$ (h) $\frac{5y}{4} \div \frac{y^2}{8}$ (i) $\frac{x^3}{z} \div \frac{x^2}{z^3}$ (j) $\frac{3e}{2f} \div \frac{12}{f^2}$.

2. Express the following fractions in their simplest form.

   (a) $\frac{3x+6}{9y}$ (b) $\frac{x^2 - 4x + 3}{3x - 9}$ (c) $\frac{x^2 + 7x + 10}{x^2 - 25}$.

3. Express the following as single fractions in their simplest form.

   (a) $\frac{2}{y+2} + \frac{4}{y-3}$, $y \neq -2$, $y \neq 3$

   (b) $\frac{1}{x+4} - \frac{1}{x+3}$, $x \neq -4$, $x \neq -3$

   (c) $\frac{4}{a+7} + \frac{3}{a}$, $a \neq -7$, $a \neq 0$

   (d) $\frac{5}{p+4} - \frac{3}{p+2}$, $p \neq -4$, $p \neq -2$

   (e) $\frac{2a}{b^2} \times \frac{b^5}{4a}$ (f) $\frac{4x}{5y} \div \frac{12}{y^2}$

   (g) $\frac{x^7}{y^3} \div \frac{x^4}{y^2}$ (h) $\frac{2b^3}{c^2} \times \frac{c}{4b^2}$.

GRADIENT

The gradient (or slope) of a straight line can be stated in the following rule.

Gradient = $\dfrac{\text{vertical height}}{\text{horizontal distance}}$

vertical height

horizontal distance

THE GRADIENT FORMULA

We can also use the **gradient formula** to find the gradient of a straight line. The letter m is normally used for gradient. The gradient formula is used to find the gradient of the straight line joining the points (x_1, y_1) and (x_2, y_2). Look at the diagram shown below.

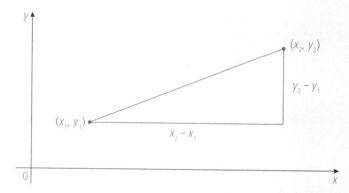

We can see that the gradient of the straight line joining the points (x_1, y_1) and (x_2, y_2) is given by the formula

$m = \dfrac{y_2 - y_1}{x_2 - x_1}$.

> **EXAMPLE:**
>
> Find the gradient of the straight line joining the points (3, – 5) and (7, 1).
>
> **SOLUTION:**
>
> $m = \dfrac{y_2 - y_1}{x_2 - x_1} = \dfrac{1 - (-5)}{7 - 3} = \dfrac{6}{4} = \dfrac{3}{2}$.

RELATIONSHIP BETWEEN GRADIENT AND EQUATION

There is a relationship between the gradient of a straight line and the equation of a straight line.

> **EXAMPLE:**
>
> Draw the straight line with equation $y = 2x$.
>
> **SOLUTION:**
>
> To draw the line, select some simple values for x, say 0, 2 and 4.
>
> Substitute them into the equation $y = 2x$ to find the corresponding values of y.
>
> This leads to the points with coordinates (0, 0), (2, 4) and (4, 8).
>
> Plot these points on a coordinate grid and join them.

contd

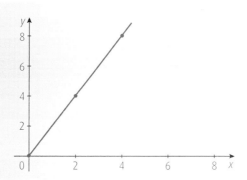

Check that the gradient of this straight line is 2, which is the coefficient of the x term in the equation $y = 2x$. We shall investigate this in more detail later.

DISTANCE-TIME GRAPHS

Gradient has a very strong link to the equation of a straight line, including such cases as the line of best fit on a scattergraph and distance–time graphs. In a distance–time graph, we shall see that the gradient of the line gives the speed. We say that speed is the **rate of change** of distance with respect to time.

EXAMPLE:

Derek travels from his home to his workplace by car and back. His journeys one day are shown in a distance–time graph.

What is the speed for each stage of the graph?

SOLUTION:

Journey to work: Speed = 50 miles per hour
At work: Speed = 0 miles per hour
Journey home: Speed = (50 ÷ 2) = 25 miles per hour.

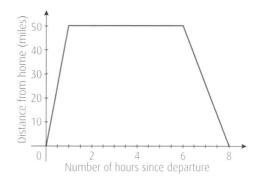

Let us now consider the gradient for the three stages using the gradient formula. We shall use the points with coordinates (0, 0), (1, 50), (6, 50) and (8, 0). We get the following results.

Journey to work: $m = \frac{y_2 - y_1}{x_2 - x_1} = \frac{50 - 0}{1 - 0} = \frac{50}{1} = 50$

At work: $m = \frac{y_2 - y_1}{x_2 - x_1} = \frac{50 - 50}{6 - 1} = \frac{0}{5} = 0$

Journey home: $m = \frac{y_2 - y_1}{x_2 - x_1} = \frac{0 - 50}{8 - 6} = \frac{-50}{2} = -25$.

We see that in a distance–time graph, the gradient of the line gives the speed.

Consider these results. We know that the speed is 50 miles per hour for the journey to work and 25 miles per hour for the journey home. We can see that a zero gradient is represented by a horizontal line. A positive gradient goes upwards from left to right, and a negative gradient goes downwards from left to right. If we were asked to find the gradient of a vertical line, the denominator in the gradient would be zero, as there is no horizontal distance. As division by zero is impossible, we say that a vertical line has an **undefined** gradient. These results for gradient are summarised below.

Positive	Negative	Zero	Undefined

ONLINE

For more on gradients, watch 'The Cartesian Plane 9: The Gradient Formula': www.brightredbooks.net/N5Maths

DON'T FORGET

Learn to use the gradient formula. You will meet it again soon.

ONLINE TEST

Test yourself on 'Gradients' at www.brightredbooks.net/N5Maths

THINGS TO DO AND THINK ABOUT

Find the gradients of the straight line joining the following pairs of points.

(a) (4, –5) and (3, 2)   (b) (–2, 2) and (3, 4)   (c) (2, 6) and (–4, 18).

ARCS OF CIRCLES

In this section, we are going to concentrate on circles. You already know that the circumference is the distance around a circle. In fact, the circumference is a special case of perimeter. You also know that the formula for the circumference of a circle is $C = \pi d$, sometimes written as $C = 2\pi r$.

EXAMPLE:

Find the circumference of a circle of radius 12 centimetres.

SOLUTION:

$C = \pi d = \pi \times 2 \times 12 = 75 \cdot 398$

Hence circumference = 75 cm (correct to 2 sig. figs).

NOTE: Unless otherwise instructed, it is good form to round the solution to the same number of significant figures as given in the original measurement. Therefore, since 12 has 2 sig. figs, round the solution to 2 sig. figs. Also, when doing circle calculations on your calculator, always use the π key.

WHAT IS AN ARC?

An **arc** of a circle is a part of the circumference. When two radii are drawn in a circle, two arcs are formed, called the major arc and the minor arc.

MAJOR ARC

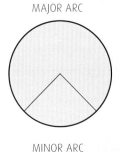

MINOR ARC

DON'T FORGET

The relationship between arc length and circumference can be used to find the length of an arc of a circle.

The length of an arc depends on the size of the angle subtended by the arc at the centre of the circle. The relationship between arc length and circumference is given below.

$$\frac{\text{arc length}}{\text{circumference}} = \frac{\text{angle at centre}}{360°}$$

EXAMPLE:

The radius of the circle, centre O, is 23 centimetres. Angle AOB = 114°. Find the length of the minor arc AB.

ONLINE

Learn more about 'Arcs' online at www. brightredbooks.net/N5Maths

SOLUTION:

Length of arc = $\frac{114}{360} \times \pi d = \frac{114}{360} \times \pi \times 2 \times 23 = 45 \cdot 76$.
Hence length of arc = 46 cm (correct to 2 sig. figs).

Suppose we are told the radius of a circle and the length of an arc in the circle and are asked to find the size of the angle at the centre of the circle. This can be done by considering the relationship given earlier in the section.

contd

EXAMPLE:

AB is an arc of a circle, centre O, with diameter 30 centimetres.
The length of arc AB is 24 centimetres.
Calculate the size of angle AOB.

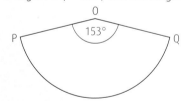

SOLUTION:

$$\frac{\text{arc length}}{\text{circumference}} = \frac{\text{angle at centre}}{360°} \Rightarrow \frac{24}{\pi \times 30} = \frac{\text{angle at centre}}{360°}$$

Hence angle at centre $= \frac{24}{\pi \times 30} \times 360 = 91\cdot673$

Angle at centre $= 92°$ (to 2 sig. figs).

ONLINE TEST

Test yourself on 'Arcs of Circles' at www.brightredbooks.net/N5Maths

 THINGS TO DO AND THINK ABOUT

1. (a) The radius of the circle, centre O, is 15 centimetres. Angle AOB = 111°.

   Find the length of the major arc AB.

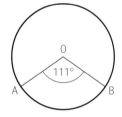

   (b) OP and OQ are radii in a circle, centre O, with diameter 35 centimetres.

   If angle POQ = 153°, find the length of arc PQ.

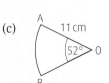

2. In the four examples below, O is the centre of each circle. Using the information given in each diagram, that is the radius of each circle and the angle at the centre of each circle, calculate the length of arc AB for each part. Give each solution correct to 2 sig. figs.

   (a)

   (b)

   (c)

   (d)

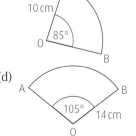

3. In a circle with centre O, the length of an arc AB is 20 centimetres.

   Calculate the size of angle AOB if the radius of the circle is 12 centimetres.

SECTORS OF CIRCLES

WHAT IS A SECTOR?

We continue our look at circles by studying sectors. A sector of a circle is an area bounded by two radii and an arc. As you will need to know how to find the area of a sector, you must use the formula for the area of a circle. You already know that the formula for the area of a circle is $A = \pi r^2$.

EXAMPLE:

Find the area of a circle of diameter 17 centimetres.

SOLUTION:

Radius = $17 \div 2 = 8{\cdot}5 \Rightarrow A = \pi \times 8{\cdot}5 \times 8{\cdot}5 = 226{\cdot}98$
Hence area = 230 cm² (correct to 2 sig. figs).

When two radii are drawn in a circle, two sectors are formed, called the major sector and the minor sector.

The area of a sector also depends on the size of the angle subtended by the arc at the centre of the circle. The relationship between the area of a sector and the area of the circle is given below.

$$\frac{\text{area of sector}}{\text{area of circle}} = \frac{\text{angle at centre}}{360°}$$

This relationship can be used to find the area of a sector of a circle.

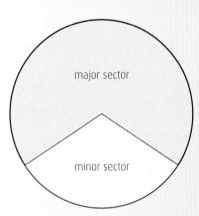

EXAMPLE:

The radius of the circle, centre O, is 11 centimetres. Angle AOB = 103°. Find the area of the minor sector OAB.

SOLUTION:

Area of sector = $\frac{103}{360} \times \pi r^2 = \frac{103}{360} \times \pi \times 11 \times 11 = 108{\cdot}76$
Hence area of sector = 110 cm² (correct to 2 sig. figs).

Suppose we are told the radius of a circle and the area of a sector in the circle and are asked to find the size of the angle at the centre of the circle. This can be done by considering the relationship given earlier in the section.

contd

EXAMPLE:

CDE is a sector of a circle, centre C, with diameter 18 centimetres.
The area of sector CDE is 92 square centimetres.
Calculate the size of angle DCE.

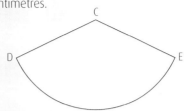

SOLUTION:

$\dfrac{\text{area of sector}}{\text{area of circle}} = \dfrac{\text{angle at centre}}{360°} \Rightarrow \dfrac{92}{\pi \times 9^2} = \dfrac{\text{angle at centre}}{360°}$

Hence angle at centre $\dfrac{92}{\pi \times 9^2} \times 360 = 130 \cdot 153$

Angle at centre = 130° (to 2 sig. figs).

A MORE DIFFICULT EXAMPLE

EXAMPLE:

In this diagram, AB and CD are arcs of circles with centres at O.

The radius OA is 12 centimetres, and the radius OC is 15 centimetres.
Angle AOB = 78°.
Calculate the shaded area.
Give your answer correct to 2 sig. figs.

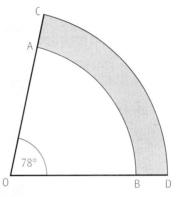

SOLUTION:

(The shaded area is the difference between the areas of two sectors.)

Area of larger sector $= \dfrac{78}{360} \times \pi \times 15^2 = 153 \cdot 152\,64$

Area of smaller sector $= \dfrac{78}{360} \times \pi \times 12^2 = 98 \cdot 017\,69$

Shaded area $= 153 \cdot 152\,64 - 98 \cdot 017\,69 = 55 \cdot 134\,95$

Hence shaded area = 55 cm² (correct to 2 sig. figs).

THINGS TO DO AND THINK ABOUT

1. (a) The radius of the circle, centre C, is 13 centimetres. Angle ACB = 98°.

   Find the area of the minor sector ACB.

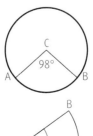

   (b) OBC is a sector of a circle, centre O, with diameter 35 centimetres.

   If angle BOC = 70°, find the length of arc BC.

2. Look back at Question 2, on the length of an arc, in the 'Things to Do and Think About' on the previous spread. For each of the four parts, calculate the area of sector OAB.

VOLUME

RECAP

You should already know the formula for the volume of a prism.

Volume of a prism = Area of base × height, or $V = Ah$ for short.

Three special types of prism are a cuboid, a cube and a cylinder.

Volume of a cuboid = length × breadth × height, or $V = lbh$ for short;

Volume of a cube = (length of side)³, or $V = l^3$ for short;

Volume of a cylinder is given by $V = \pi r^2 h$.

You should also remember the basic units of volume, including the fact that 1 litre = 1000 cubic centimetres (1 l = 1000 cm³ for short).

VOLUMES OF STANDARD SOLIDS

Next, we shall look at the volumes of other standard solids such as pyramids, cones and spheres.

The formula for the volume of a pyramid is $V = \frac{1}{3}Ah$ where A is the area of the base and h is the perpendicular height.

EXAMPLE:

Find the volume of the square pyramid shown.

The pyramid has perpendicular height 15 centimetres and a square base of side 8 centimetres.

SOLUTION:

$V = \frac{1}{3}Ah = \frac{1}{3} \times 8 \times 8 \times 15 = 320$

Volume of pyramid = 320 cm³.

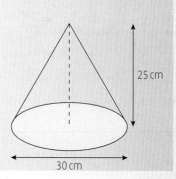

A special type of pyramid, with a circular base, is a cone. The formula for the volume of a cone is $V = \frac{1}{3}\pi r^2 h$.

EXAMPLE:

A cone has height 25 centimetres and diameter 30 centimetres.

Calculate the volume of the cone.

Give your answer correct to 3 sig. figs.

SOLUTION:

$V = \frac{1}{3}\pi r^2 h = \frac{1}{3} \times \pi \times 15^2 \times 25 = 5890{\cdot}486\,225$

Hence volume = 5890 cm³ (to 3 sig. figs).

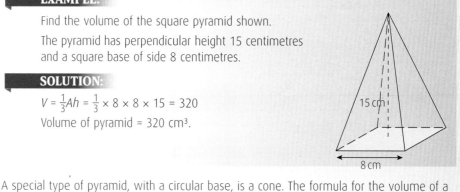

We shall now look at spheres. The formula for the volume of a sphere is $V = \frac{4}{3}\pi r^3$.

EXAMPLE:

A sphere has diameter 14 centimetres.
Calculate its volume.
Give your answer correct to 3 sig. figs.

ONLINE

Have a look at the site with 'Volume formulas' at www.brightredbooks.net/N5Maths. How many do you know?

DON'T FORGET

Always use the π key on your calculator. Notice that the diameter has to be halved to use in the formula. You should also know how to use your calculator to do 15^2 without doing 15 × 15. Remember to round your solution at the end.

contd

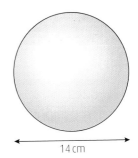

14 cm

SOLUTION:

$V = \frac{4}{3}\pi r^3 = \frac{4}{3} \times \pi \times 7^3 = 1436 \cdot 755\,04$
Hence volume = 1440 cm³ (to 3 sig. figs).

NOTE: You should also know how to use your calculator to do 7^3 without doing $7 \times 7 \times 7$.

WORKING BACK

On some occasions, you will know the volume of a standard solid such as a cylinder, cone or sphere. You may have to work back to find a missing dimension in the solid, for example the height or the radius. Remember that the volume formulae for such solids involve multiplication, so, when you are working back, you will have to use the opposite operation, namely division. Study the following examples to see how this can be done.

EXAMPLE:

The volume of a cone is 1500 cubic centimetres.
Its height is 8 centimetres. Calculate its radius.

SOLUTION:

Write down the formula for the volume of a cone: $V = \frac{1}{3}\pi r^2 h$.

Now substitute into the formula, that is $1500 = \frac{1}{3} \times \pi \times r^2 \times 8$.

Now start to find r^2 by division, that is $r^2 = \frac{1500}{\frac{1}{3} \times \pi \times 8} = \frac{1500}{8 \cdot 377\,580\,41} = 179 \cdot 049$.

To find r, take the square root of r^2, that is $r = \sqrt{179 \cdot 049} = 13 \cdot 38$.

Hence the radius is 13 cm (correct to 2 sig. figs).

NOTE: When you have studied 'Changing the subject of a formula' later in the course (p. 66), you will find a different approach to this type of problem. However, the method shown above is valid. We will return to this example later, and you can decide which approach you favour then.

If we have to carry out a working-back problem with a sphere, this is trickier. Suppose we know the volume of a sphere and want to find its radius. The formula for the volume of a sphere is $V = \frac{4}{3}\pi r^3$, therefore if we are working back, we will have to take the **cube root** to find the solution. Remember that the symbol for cube root is $\sqrt[3]{}$.

EXAMPLE:

A sphere has volume 5000 cubic centimetres. Find its radius.

SOLUTION:

$V = \frac{4}{3}\pi r^3 \Rightarrow 5000 = \frac{4}{3}\pi r^3 \Rightarrow r^3 = \frac{5000}{\frac{4}{3} \times \pi} = \frac{5000}{4 \cdot 188\,79} = 1193 \cdot 66$

$r^3 = 1193 \cdot 66 \Rightarrow r = \sqrt[3]{1193 \cdot 66} = 10 \cdot 607\,8$

Hence the radius = 10·6 cm (correct to 3 sig. figs).

 DON'T FORGET

If you are working back to find the radius of a sphere, you need to calculate a cube root, so make sure you know how to do this on your calculator.

 ONLINE TEST

Test yourself on 'Volume' at www.brightredbooks.net/ N5Maths

 THINGS TO DO AND THINK ABOUT

(Give all answers correct to 3 sig. figs)

1. Find the volume of a cylinder with radius 10 cm and height 35 cm.

2. Find the volume of a pyramid with a square base of side 9 cm and height 12 cm.

3. Find the volume of a cone with diameter 22 cm and height 19 cm.

4. Find the volume of a sphere with diameter 16 cm.

5. Find the height of a cylinder with volume 2500 cubic cm and radius 10 cm.

MORE VOLUME

We shall now look at some more complicated problems on volume. These will include composite shapes. A composite shape is made up of different parts, for example a cylinder with a cone on top. In such cases, the volume will be calculated by addition.

In other situations, the volume of a shape may have to be calculated by subtracting volumes. For example, the volume of a pipe may be found by subtracting the volume of one cylinder from another.

We shall look in some detail at different types of volume problems in this section.

ADDING VOLUMES

EXAMPLE:

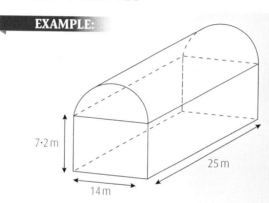

7·2 m

25 m

14 m

A barn shaped like a cuboid with a semi-cylindrical roof has dimensions as shown in the diagram.

Calculate the volume of the barn in cubic metres, giving your answer correct to 3 sig. figs.

SOLUTION:

Volume of barn = Volume of cuboid + Volume of semi-cylinder

Volume of cuboid: $V = lbh = 25 \times 14 \times 7\cdot2 = 2520$

Volume of semi-cylinder: $V = \frac{1}{2}\pi r^2 h = \frac{1}{2} \times \pi \times 7^2 \times 25 = 1924\cdot225\,5$

Volume of barn = $2520 + 1924\cdot225\,5 = 4444\cdot225\,5$

Hence volume of barn = 4440 m³ (to 3 sig. figs).

NOTE: This problem could also be solved using the formula for the volume of a prism, $V = Ah$. Calculate A, the area of the end of the barn (a rectangle + a semi-circle), then multiply by h, the length of the barn. This method is equally valid.

Check all the working carefully, and remember not to round until the end.

SUBTRACTING VOLUMES

EXAMPLE:

(a) An ice-cream tub is in the shape of part of a cone with dimensions as shown.

Calculate the volume of the ice-cream tub.

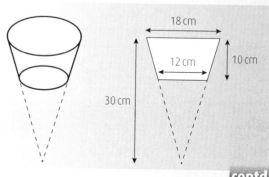

18 cm

12 cm

10 cm

30 cm

contd

(b) Another ice-cream tub is designed in the shape of a cylinder.
It has the **same** volume as the first tub.
The diameter of this tub is 15 centimetres.
Calculate the height of the tub.

15 cm

SOLUTION:

(a) Volume of ice-cream tub = Volume of larger cone – Volume of smaller cone
Volume of larger cone: $V = \frac{1}{3}\pi r^2 h = \frac{1}{3} \times \pi \times 9^2 \times 30 = 2544{\cdot}690\,0$
Volume of smaller cone: $V = \frac{1}{3}\pi r^2 h = \frac{1}{3} \times \pi \times 6^2 \times (30 - 10) = 753{\cdot}982\,2$
Volume of ice-cream tub = $2544{\cdot}690\,0 - 753{\cdot}982\,2 = 1790{\cdot}707\,8$
Hence volume of tub = 1791 cm³.

(b) Volume of cylindrical tub: $V = \pi r^2 h \Rightarrow 1791 = \pi \times 7{\cdot}5^2 \times h$
Hence $h = \frac{1791}{\pi \times 7{\cdot}5^2} = \frac{1791}{177} = 10{\cdot}1$
Hence height is 10·1 cm.

It is possible you could be asked a question on volume without the use of a calculator. In this instance, you would be given the approximate value for π that is 3·14.

VOLUME WITHOUT A CALCULATOR

EXAMPLE:

A sphere has diameter 6 centimetres.
Find its volume. Take $\pi = 3{\cdot}14$.

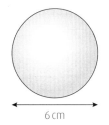

6 cm

SOLUTION:

$V = \frac{4}{3}\pi r^3 = \frac{4}{3} \times 3{\cdot}14 \times 3^3 = \frac{4}{3} \times 3{\cdot}14 \times 3 \times 3 \times 3 = 113{\cdot}04$
Volume = 113·04 cm³.

Most students will find this calculation tricky without a calculator, but it is manageable if you approach it in the most efficient way. First, you must replace π with 3·14 and write 3^3 as $3 \times 3 \times 3$. Then remove the fraction (the cause of most of the difficulties) by cancelling it with the final 3 on the numerator, leaving $4 \times 3{\cdot}14 \times 3 \times 3$. Now multiply $4 \times 3{\cdot}14 = 12{\cdot}56$, then $12{\cdot}56 \times 3 = 37{\cdot}68$ and finally $37{\cdot}68 \times 3 = 113{\cdot}04$. To sum up, if you remove the fraction by cancelling, you can then do three fairly simple multiplications by 4, then 3, then 3 again. Do not round your answer this time, to show that you carried out all the calculations correctly.

THINGS TO DO AND THINK ABOUT

A petrol container is in the shape of a cylinder with hemispherical ends, as shown in the diagram. The total length of the tank is 1·8 metres, and the length of the cylinder is 1·3 metres.

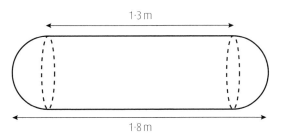

1·3 m

1·8 m

Calculate the volume of the tank in cubic metres.

Give your answer correct to 2 sig. figs.

THE EQUATION OF A STRAIGHT LINE

DRAWING A STRAIGHT LINE

In the earlier section on Gradient (p. 36), we looked at how to draw the straight line with equation $y = 2x$.

When we are given the equation of a straight line, we can draw the line by choosing some values for one variable (usually x) and substituting them into the equation to find the corresponding values of y. By then plotting the points (three would be recommended), we can draw the line. We can also check whether a given point lies on the straight line or not.

> **EXAMPLE:**
>
> Does the point (3, 5) lie on the straight line with equation $3x - 2y = 1$?
>
> **SOLUTION:**
>
> We substitute $x = 3$ and $y = 5$ into the left-hand side of the equation of the line.
> $3x - 2y \Rightarrow 3 \times 3 - 2 \times 5 = 9 - 10 = -1$. If it was on the line, we would get 1.
> Hence (3, 5) does not lie on the line $3x - 2y = 1$.

GRADIENT

We also looked at two ways of finding the gradient of a straight line.

The gradient (or slope) of a straight line can be stated in the following rule.

$$\text{Gradient} = \frac{\text{vertical height}}{\text{horizontal distance}}$$

vertical height

horizontal distance

We can also use the **gradient formula**.

The gradient of the straight line joining the points (x_1, y_1) and (x_2, y_2) is given by the formula $m = \frac{y_2 - y_1}{x_2 - x_1}$.

You should also remember the facts about positive, negative, zero and undefined gradients summarised in the table below.

Positive	Negative	Zero	Undefined

THE EQUATION $y = mx + c$

Now we will investigate how to find the equation of a straight line. The formula for the equation of a straight line is $y = mx + c$, where m is the gradient of the straight line and c is the y-intercept.

The y-intercept is the y-coordinate of the point where the straight line cuts the y-axis. In other words, if a straight line cuts the y-axis at $(0, c)$, then c is the y-intercept.

Note that the equation $y = mx$ refers to a straight line with gradient m which passes through the origin as $c = 0$.

contd

EXAMPLE:

Find the equation of the straight line shown.

SOLUTION:

To find the gradient, select two suitable points on the straight line, for example (0, 8) and (4, 0).

Then find gradient. If you use the first method, note that the line has a negative gradient:

$m = \dfrac{\text{vertical height}}{\text{horizontal distance}} = -\dfrac{8}{4} = -2$

or $m = \dfrac{y_2 - y_1}{x_2 - x_1} = \dfrac{0 - 8}{4 - 0} = \dfrac{-8}{4} = -2$.

As the line crosses the y-axis at (0, 8), the y-intercept is 8, hence $c = 8$.

Using the formula $y = mx + c$, the equation of the straight line is $y = -2x + 8$.

STRAIGHT LINES PARALLEL TO THE X- AND Y-AXES

An equation of the form $y = k$ where k is a constant is the equation of a straight line parallel to the x-axis. It represents a horizontal line with gradient $m = 0$ and it crosses the y-axis at the point (0, k).

An equation of the form $x = h$ where h is a constant is the equation of a straight line parallel to the y-axis. It represents a vertical line with an undefined gradient. Note that if you use the gradient formula on a line of this type, the denominator would be zero – and it is impossible to divide by zero.

EXAMPLE:

On the same grid, draw the straight lines with equations $x = 4$ and $y = 2$.

SOLUTION:

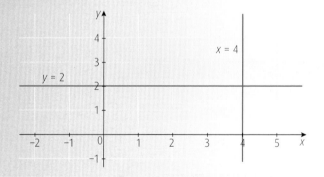

 ## THINGS TO DO AND THINK ABOUT

Find the equation of the straight line shown in the diagram.

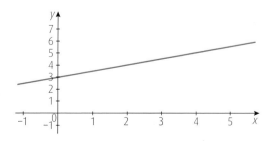

MORE ON THE STRAIGHT LINE

CROSSING THE AXES

You should be aware that equations such as $y = \frac{2}{3}x + 6$, $3y - 2x = 18$ and $2x - 3y + 18 = 0$ all represent the same straight line. By removing fractions and rearranging terms, the equation of a straight line can appear in different forms. Consider where the above line crosses the x- and y-axes.

> **EXAMPLE:**
>
> Find the coordinates of the points where the straight line $3y - 2x = 18$ crosses the x- and y-axes.
>
> **SOLUTION:**
>
> The line crosses the x-axis where $y = 0$.
> Hence $3y - 2x = 18 \Rightarrow (3 \times 0) - 2x = 18 \Rightarrow -2x = 18 \Rightarrow x = -9$
> The line crosses the y-axis where $x = 0$.
> Hence $3y - 2x = 18 \Rightarrow 3y - (2 \times 0) = 18 \Rightarrow 3y = 18 \Rightarrow y = 6$
> The line crosses the axes at $(-9, 0)$ and $(0, 6)$.
> NOTE: Finding where a line crosses the axes can be useful if you have to draw the line.

REARRANGING THE EQUATION OF A STRAIGHT LINE

We have noticed that the equation of the same straight line can appear in different forms. By rearranging the equation into the form $y = mx + c$, it is easy to find the gradient and y-intercept.

> **EXAMPLE:**
>
> Find the gradient of the straight line with equation $5y + 2x = 20$.
>
> **SOLUTION:**
>
> $5y + 2x = 20 \Rightarrow 5y = -2x + 20 \Rightarrow y = -\frac{2}{5}x + \frac{20}{5} \Rightarrow y = -\frac{2}{5}x + 4$
> Hence the gradient $m = -\frac{2}{5}$.
> NOTE: We can also see from the rearrangement that the y-intercept is 4.

ANOTHER VERSION OF THE EQUATION OF A STRAIGHT LINE

We have seen that the formula $y = mx + c$ is particularly useful for finding the equation of a straight line when we are given a suitable diagram and when it is clear that the y-intercept is an integer. However, in other situations, a different formula is more suitable.

Suppose a straight line with gradient m passes through the point with coordinates (a, b).

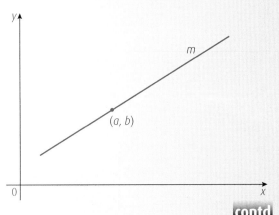

contd

The equation of this line is given by the formula $y - b = m(x - a)$.

This formula will prove very useful in finding the equation of a straight line, and it is very important that you practise using it until you are confident. Some textbooks give this formula in a slightly different notation, that is $y - y_1 = m(x - x_1)$.

> **EXAMPLE:**
>
> Find the equation of the straight line with gradient 2 passing through the point with coordinates (3, 5).
>
> **SOLUTION:**
>
> Equation is $y - b = m(x - a) \Rightarrow y - 5 = 2(x - 3) \Rightarrow y - 5 = 2x - 6$
> $$\Rightarrow y = 2x - 1.$$
>
> NOTE: The solution has been written in the form $y = mx + c$. By doing this, it is easy to see that the gradient is 2 and to check that (3, 5) lies on the line.

> **EXAMPLE:**
>
> Find the equation of the line passing through the points (4, –2) and (–6, 3).
>
> **SOLUTION:**
>
> As we will need to use the formula $y - b = m(x - a)$, we first find the gradient.
> $m = \frac{y_2 - y_1}{x_2 - x_1} = \frac{3 - (-2)}{-6 - 4} = \frac{5}{-10} = -\frac{1}{2}$.
> Equation is $y - b = m(x - a) \Rightarrow y - (-2) = -\frac{1}{2}(x - 4) \Rightarrow y + 2 = -\frac{1}{2}(x - 4)$.
> Note that we can use either point for (a, b). Here, (4, –2) was chosen.
> The solution given is the equation of the line but would normally be simplified.
> Double both sides to remove the fraction, then multiply out the brackets.
> $y + 2 = -\frac{1}{2}(x - 4) \Rightarrow 2(y + 2) = -1(x - 4) \Rightarrow 2y + 4 = -x + 4 \Rightarrow x + 2y = 0$.
> NOTE: We have seen already that the equation of a line can be written in different ways by rearranging the terms. The final solution above could have been written in different ways. The solution $x + 2y = 0$ seems appropriate as there are no negative signs involved; however, other versions are equally valid.

ONLINE TEST

Take the 'More on the Straight Line' test online at www.brightredbooks.net/N5Maths

THINGS TO DO AND THINK ABOUT

To summarise the work on the straight line so far, we can find the equation of a straight line using the formulae $y = mx + c$ or $y - b = m(x - a)$. The first formula is useful when you are given a diagram where it is easy to find the gradient and the y-intercept. The second formula works in every case, easy or difficult. Similarly, the gradient formula works in every case and is recommended.

The formula $y = mx + c$ is very useful for finding the gradient and y-intercept of a line. You may have to rearrange an equation first. Remember too how to find where a straight line crosses the x- and y-axes.

1. Find where the line $4x + 3y = 12$ crosses the x- and y-axes.

2. Find the gradient of the line $6x + 2y = 5$.

3. Find the gradient of the straight line joining the points (–1, 3) and (1, 7).

4. Find the equation of the straight line with gradient 4 passing through the point (0, 6).

5. Find the equation of the straight line with gradient –2 passing through the point (3, 4).

6. Find the equation of the straight line joining the points (–1, 3) and (8, 3).

7. Find the equation of the straight line joining the points (2, 4) and (5, 13).

STRAIGHT-LINE PROBLEMS

PROBLEMS IN CONTEXT

Problems on the straight line can often occur in context. Later, we will investigate how to find the equation of the line of best fit in a scattergraph. In the meantime, study the following example.

EXAMPLE:

Tony buys a 10-litre bag of cat litter.
Each day, his cat uses 400 millilitres of cat litter.
The graph below shows the volume of cat litter remaining in the bag (V litres) against time (d days).

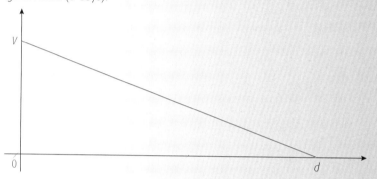

Find an equation connecting V and d.

SOLUTION:

We are really being asked to find the equation of the straight line in the diagram.

So, we must think of this problem as one of finding the equation of a straight line. After 0 days, the volume is 10 litres, so (0, 10) lies on the line. The bag will last for (10 ÷ 0·4) days, that is 25 days, so (25, 0) lies on the line.

Gradient is $m = \frac{y_2 - y_1}{x_2 - x_1} = \frac{0 - 10}{25 - 0} = \frac{-10}{25} = -\frac{2}{5}$.

y-intercept is 10.

Hence, using the formula $y = mx + c$, the equation of the line is $y = -\frac{2}{5}x + 10$.

Hence the equation connecting V and d is $V = -\frac{2}{5}d + 10$.

NOTE: It is important that x and y are replaced with V and d.

VIDEO LINK

Watch a short explanation by clicking on 'Parallel lines – Gradient: Exam Solutions' at www.brightredbooks.net/N5Maths

PARALLEL LINES

Gradient is a measure of slope and, as parallel lines have the same slope, we can see that parallel lines have the same gradient.

EXAMPLE:

The line joining A (2, 1) to B (6, 9) is parallel to the line joining C (3, 6) to D (7, t). Find the value of t.

SOLUTION:

The gradients of lines AB and CD can be denoted by m_{AB} and m_{CD} respectively.

$m_{AB} = \frac{y_2 - y_1}{x_2 - x_1} = \frac{9 - 1}{6 - 2} = \frac{8}{4} = 2$ and $m_{CD} = \frac{y_2 - y_1}{x_2 - x_1} = \frac{t - 6}{7 - 3} = \frac{t - 6}{4}$

Hence $\frac{t - 6}{4} = 2 \Rightarrow t - 6 = 8 \Rightarrow t = 14$.

DON'T FORGET

Parallel lines have the same gradient.

contd

EXAMPLE:

Find the equation of the straight line passing through the point (4, –3) which is parallel to the straight line with equation $2x + y = 16$.

SOLUTION:

As parallel lines have the same gradient, we must find the gradient of the line $2x + y = 16$. To do this, rearrange the equation into the form $y = mx + c$.

$2x + y = 16 \Rightarrow y = -2x + 16$, hence the required gradient is –2.

Now find the equation of the parallel line using the formula $y - b = m(x - a)$.

$y - b = m(x - a) \Rightarrow y - (-3) = -2(x - 4) \Rightarrow y + 3 = -2x + 8 \Rightarrow y = -2x + 5$.

SUMMARY

After studying the straight line, you should be able to

- Draw a straight line given its equation
- Find the gradient of a straight line using gradient = $\dfrac{\text{vertical height}}{\text{horizontal distance}}$
- Find the gradient of a straight line using the gradient formula $m = \dfrac{y_2 - y_1}{x_2 - x_1}$
- Find the equation of a straight line using the formula $y = mx + c$
- Find the equation of a straight line using the formula $y - b = m(x - a)$
- Understand the equations of straight lines parallel to the x- and y-axes
- Rearrange the equation of a straight line where necessary
- Solve problems on the straight line in context
- Solve problems involving parallel gradients.

Clearly, the topic of the straight line has many different areas of study. You will encounter the straight line in several future topics, so it is extremely important that you learn the formulae for gradient and the equation of a straight line and practise using them until you feel confident.

THINGS TO DO AND THINK ABOUT

ONLINE TEST

Take the 'Straight Line Problems' test online at www.brightredbooks.net/N5Maths

1. A bath contains 120 litres of water. When the plug is removed from the bath, water is emptied at the rate of 20 litres per minute.

   The graph shows the volume of water remaining in the bath (V litres) against time (t minutes)

   Find an equation connecting V and t.

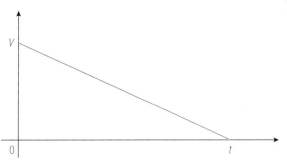

2. Find the equation of the straight line passing through the point (–1, –2) which is parallel to the straight line with equation $y - 5x = 4$.

3. P and Q are the points (–3, 0) and (7, 5) respectively. RS is a straight line which is parallel to PQ. If the coordinates of R are (2, 6), find the equation of RS.

FUNCTIONAL NOTATION

WHAT IS A FUNCTION?

In mathematics, a **function** is a relationship between two sets such that each member of the first set is related to exactly one member of the second set. An example is the function which relates each real number to double the number.

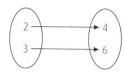

Each real number x would be related to double the number, that is $2x$.

We therefore get pairs of numbers: (2, 4), (3, 6) and so on. By joining these coordinates, we can draw the graph of the function. This is exactly what we did earlier when we drew a sketch of the straight line with equation $y = 2x$.

However, another notation is sometimes used for a function. It is called functional notation and is based on the work of a famous German mathematician called Gottfried Wilhelm von Leibnitz, who lived from 1646 to 1716.

The functional notation $y = f(x)$ is based on his work.

We say 'f of x' when we read $f(x)$.

In functional notation, we would say that $f(x) = 2x$.

This means the same as $y = 2x$.

EXAMPLE:

Draw the graph represented by the function $f(x) = x + 3$.

SOLUTION:

$f(x) = x + 3$ can be thought of as $y = x + 3$ and treated as the equation of a straight line.

Pick some simple values for x, say 0, 2 and 4, and substitute them into $y = x + 3$ to find the corresponding values of y leading to the points (0, 3), (2, 5) and (4, 7). This information is sometimes put in a table:

x	0	2	4
$f(x) = x + 3$	3	5	7

By plotting the points, we get the graph of $f(x) = x + 3$.

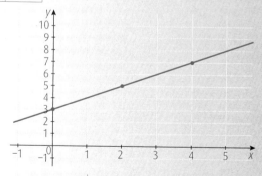

The above function is called a linear function because its graph is a straight line. Functions of the form $f(x) = ax + b$ are linear functions. We will also study functions of the type $f(x) = ax^2 + bx + c$, which are called quadratic functions. Their graphs, mentioned earlier, are called parabolas.

USING FUNCTIONAL NOTATION

When we are given a function in the form $y = f(x)$, we can find the value of the function for different values of x by substitution.

EXAMPLE:

If $f(x) = 6 - 5x$, evaluate $f(-2)$.

SOLUTION:

Simply replace x by -2 in the function.

Hence $f(-2) = 6 - 5 \times (-2) = 6 + 10 = 16$.

NOTE: We say that 16 is the image of -2. Note also that the point with coordinates $(-2, 16)$ would lie on the graph of $y = 6 - 5x$. Remember the correct order of operations, that is multiplication/division before addition/subtraction.

LOOKING BACK AND LOOKING AHEAD

In the earlier section on 'Completing the square' (p. 30), we looked at a parabola based on the quadratic equation $y = x^2 - 12x + 25$. By considering the quadratic function

$$f(x) = x^2 - 12x + 25$$

and completing a table of values, you can see how the graph was formed.

x	2	3	4	5	6	7	8	9	10
x^2	4	9	16	25	36	49	64	81	100
$-12x$	-24	-36	-48	-60	-72	-84	-96	-108	-120
$+25$	$+25$	$+25$	$+25$	$+25$	$+25$	$+25$	$+25$	$+25$	$+25$
$f(x) = x^2 - 12x + 25$	5	-2	-7	-10	-11	-10	-7	-2	5

Values of x from 2 to 10 have been substituted into $f(x) = x^2 - 12x + 25$. To ease working, the table has been split into the three component parts of the function. The points from the table $(2, 5)$, $(3, -2)$, up to $(10, 5)$ all lie on the parabola. Notice the symmetry of the graph.

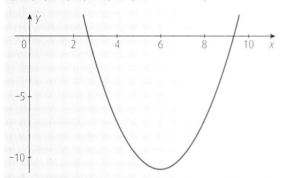

Note also that the minimum turning point is $(6, -11)$. We discovered this earlier through completing the square.

We shall study quadratic functions in more detail later.

ONLINE TEST

Take the 'Functional Notation' test online at www.brightredbooks.net/N5Maths

THINGS TO DO AND THINK ABOUT

1. If $f(x) = 8x - 3$, evaluate $f(2)$.

2. If $f(x) = x^2 + 2x - 5$, find the value of $f(3)$.

3. If $f(x) = x^2 - 4x + 2$, find the value of $f(-4)$.

LINEAR EQUATIONS

LOOKING AT EQUATIONS

Equations form a vital part of your studies in algebra. An equation is a sentence with the verb phrase 'is equal to' in it. A linear equation is a straightforward equation in which the terms are either constants or the product of a constant and a variable, for example $3x + 5 = 11$. The process of finding the value of the variable x is known as solving the equation.

We have already come across some equations in earlier sections. In this section, we shall concentrate on equations whose coefficients are integers. The set of integers consists of all positive and negative whole numbers and zero, that is $\{... -3, -2, -1, 0, 1, 2, 3 ...\}$. The set of integers is represented by the letter Z.

REMINDERS

EXAMPLE:

Consider the above example $3x + 5 = 11$. It is often thought of as a balancing type of problem in which a set of scales can only be kept balanced by doing the same thing to both sides.

Hence we would start by subtracting 5 from each side of the equation. The whole working process might be set out as shown below.

SOLUTION:

$3x + 5 = 11$
$\quad - 5 \quad - 5$ (subtract 5 from each side of the equation)
$3x \quad = 6$
$\quad \div 3 \quad \div 3$ (divide each side of the equation by 3)
$x \quad = 2.$

The working for the above equation is often set out in a more compact form:
$3x + 5 = 11$
$\Rightarrow 3x = 11 - 5$
$\Rightarrow 3x = 6$
$\Rightarrow x = 2.$

 DON'T FORGET

When solving a linear equation, you should aim to end up with the variable terms on one side and the constant terms on the other side. Whatever you do to one side, that is add, subtract, multiply or divide, you must do the same to the other side to keep things balanced.

 DON'T FORGET

If you are asked to solve an equation algebraically, it is essential that you show working of the type illustrated alongside, or equivalent. For example, in part (a), writing down that $3 \times 6 - 7 = 11$, therefore $y = 6$ would be unacceptable.

EXAMPLE:

Solve algebraically the following equations.
(a) $3y - 7 = 11$ (b) $8 + 4x = 24$ (c) $12 - 2x = 10.$

SOLUTION:

(a) $3y - 7 = 11 \Rightarrow 3y = 11 + 7 \Rightarrow 3y = 18 \Rightarrow y = 6$
(b) $8 + 4x = 24 \Rightarrow 4x = 24 - 8 \Rightarrow 4x = 16 \Rightarrow x = 4$
(c) $12 - 2x = 10 \Rightarrow -2x = 10 - 12 \Rightarrow -2x = -2 \Rightarrow x = 1.$

MORE DIFFICULT EQUATIONS

We shall now look at equations where there are variables and constants on both sides.

EXAMPLE:

Solve algebraically the equation $5x - 7 = 2x + 20.$

contd

SOLUTION:

$5x - 7 = 2x + 20 \Rightarrow 5x - 2x = 20 + 7 \Rightarrow 3x = 27 \Rightarrow x = 9.$

Check that you follow this compact way of setting out the working. Basically, we have added 7 to both sides and subtracted $2x$ from both sides, leading to $3x = 27$. Finally, we divide both sides by 3 for the solution.

EQUATIONS WITH RATIONAL SOLUTIONS

Often, the solution to an equation will not be an integer but a rational number. Remember that a rational number is a number which can be written as a fraction. If the fraction can be simplified, you should do so.

EXAMPLE:

Solve algebraically the equation $5x + 1 = x + 19$.

SOLUTION:

$5x + 1 = x + 19 \Rightarrow 5x - x = 19 - 1 \Rightarrow 4x = 18 \Rightarrow x = \frac{18}{4} = \frac{9}{2}.$

NOTE: It is acceptable to leave the solution in the form $\frac{9}{2}$.

EQUATIONS WITH BRACKETS

If you encounter a linear equation involving brackets, you should multiply out the brackets first, then solve as shown above.

EXAMPLE:

Solve algebraically the following equations.
(a) $3(y + 4) = 30$ (b) $4(2y + 3) = 3(y + 1)$.

SOLUTION:

(a) $3(y + 4) = 30 \Rightarrow 3y + 12 = 30 \Rightarrow 3y = 30 - 12 \Rightarrow 3y = 18 \Rightarrow y = 6$

(b) $4(2y + 3) = 3(y + 1) \Rightarrow 8y + 12 = 3y + 3 \Rightarrow 8y - 3y = 3 - 12 \Rightarrow 5y = -9 \Rightarrow y = -\frac{9}{5}.$

> **BEWARE: Many of the errors which occur in solving equations are caused by wrong signs: that is, + and – signs are confused. Study the next example carefully, paying particular attention to the signs.**

EXAMPLE:

Solve algebraically the equation $4(3 - 2a) = 2(a - 4)$.

SOLUTION:

$4(3 - 2a) = 2(a - 4) \Rightarrow 12 - 8a = 2a - 8 \Rightarrow 12 + 8 = 2a + 8a \Rightarrow 20 = 10a$
$$\Rightarrow a = 2.$$

In this case, the variables have been placed on the right side and the constants on the left side. This is probably preferable, as it avoids a lot of negative terms. Always choose the simplest way to solve an equation.

DON'T FORGET

There are several different strategies for setting out working when solving linear equations, although all are essentially doing the same thing. Some students use the balancing ideas shown at the start of the section, some use a more compact form of this, and some use mnemonics such as 'Change the side, change the sign'. Whatever method is employed, as long as it is algebraic and leads you to consistently get the correct solution, then there will not be a problem.

VIDEO LINK

Click 'Solving equations with brackets easily' at www.brightredbooks.net/N5Maths to see an example that solves an equation with brackets and the balancing approach.

ONLINE TEST

Take the 'Linear Equations' test online at www.brightredbooks.net/N5Maths

THINGS TO DO AND THINK ABOUT

Solve algebraically the following equations.

(a) $6x + 5 = 4x + 17$ (b) $4x - 3 = x + 12$ (c) $10 - 6b = 4b + 40$

(d) $5y - 2 = 3y - 7$ (e) $7(u + 3) = 2(2u + 9)$ (f) $4(2 - 3a) = 6(1 - a)$.

FURTHER EQUATIONS

EQUATIONS INVOLVING FRACTIONS

The introduction of fractions into algebraic equations always causes extra difficulty.

> **EXAMPLE:**
>
> Solve the equation $\frac{2x}{5} = \frac{3}{4}$.
>
> **SOLUTION:**
>
> The first step would be to remove the main difficulty, that is, get rid of the fractions. This can be done by multiplying both sides of the equation by the LCM (Least Common Multiple) of 5 and 4, the denominators of the fractions. The LCM of 5 and 4 is 20.
>
> Hence $\frac{2x}{5} = \frac{3}{4} \Rightarrow 20 \times \frac{2x}{5} = 20 \times \frac{3}{4} \Rightarrow \frac{40x}{5} = \frac{60}{4} \Rightarrow 8x = 15 \Rightarrow x = \frac{15}{8}$.
>
> We can see that when we multiply both sides by 20, the denominators of the fractions (5 and 4) divide into the numerators exactly, and this gets rid of the fractions. The equation then becomes much simpler to solve. However there is a shorter and neater way to solve the equation.

CROSS-MULTIPLICATION

We can solve equations involving two equal fractions more quickly by cross-multiplication.

$$\frac{2x}{5} \diagdown = \diagup \frac{3}{4}$$

Simply multiply the denominator of each fraction by the numerator of the opposite fraction, leading to $\frac{2x}{5} = \frac{3}{4} \Rightarrow 2x \times 4 = 3 \times 5 \Rightarrow 8x = 15 \Rightarrow x = \frac{15}{8}$.

Clearly, this is quick and efficient.

> **EXAMPLE:**
>
> Solve the equation $\frac{12}{5x} = \frac{2}{3}$.
>
> **SOLUTION:**
>
> $\frac{12}{5x} = \frac{2}{3} \Rightarrow 12 \times 3 = 2 \times 5x \Rightarrow 36 = 10x \Rightarrow x = \frac{36}{10} = \frac{18}{5}$.
>
> The technique of cross-multiplication can be used to solve equations of the type $\frac{x}{3} = 4$ (by thinking of it as $\frac{x}{3} = \frac{4}{1}$) and of the type $\frac{2}{5}x = \frac{3}{7}$ (by thinking of it as $\frac{2x}{5} = \frac{3}{7}$).
>
> Check that the solutions of these equations are $x = 12$ and $x = \frac{15}{14}$ respectively.

> **EXAMPLE:**
>
> Solve the equation $\frac{x+3}{2} = \frac{x-2}{5}$.
>
> **SOLUTION:**
>
> $\frac{x+3}{2} = \frac{x-2}{5} \Rightarrow 5(x+3) = 2(x-2) \Rightarrow 5x + 15 = 2x - 4$
> $\Rightarrow 5x - 2x = -4 - 15 \Rightarrow 3x = -19 \Rightarrow x = -\frac{19}{3}$.

> **EXAMPLE:**
>
> Solve the equation $\frac{x}{4} = \frac{1}{2} + \frac{x}{3}$.
>
> **SOLUTION:**
>
> We cannot cross–multiply here, as we do not have two equal fractions. There are

contd

ONLINE

Check 'Fractions & Proportions: How to cross multiply proportions by eHow' at www. brightredbooks.net/N5Maths

---

DON'T FORGET

Cross-multiplication is a quick method for solving equations which involve two equal fractions.

BEWARE: Not every equation involving fractions can be solved by cross-multiplication. Consider the following example.

---

two terms added together on the right-hand side of the equation. We still should get rid of the fractions, however. Multiply throughout by the LCM of 2, 3 and 4, that is, 12.

Hence $\frac{x}{4} = \frac{1}{2} + \frac{x}{3} \Rightarrow 12 \times \frac{x}{4} = 12 \times \frac{1}{2} + 12 \times \frac{x}{3} \Rightarrow \frac{12x}{4} = \frac{12}{2} + \frac{12x}{3} \Rightarrow 3x = 6 + 4x$.

Once the fractions have been removed, it is easier to solve.

$3x = 6 + 4x \Rightarrow 3x - 4x = 6 \Rightarrow -x = 6 \Rightarrow x = -6$.

A MORE DIFFICULT EXAMPLE

EXAMPLE:

Solve the equation $\frac{1}{2}(y - 2) + \frac{3}{4}(3 - y) = \frac{1}{8}(y + 1)$.

SOLUTION:

Multiply throughout by the LCM of 2, 4 and 8, which is 8.

$\frac{1}{2}(y - 2) + \frac{3}{4}(3 - y) = \frac{1}{8}(y + 1) \Rightarrow 8 \times \frac{1}{2}(y - 2) + 8 \times \frac{3}{4}(3 - y) = 8 \times \frac{1}{8}(y + 1)$

$\Rightarrow 4(y - 2) + 6(3 - y) = 1(y + 1)$

$\Rightarrow 4y - 8 + 18 - 6y = y + 1$

$\Rightarrow 4y - 6y - y = 1 + 8 - 18 \Rightarrow -3y = -9 \Rightarrow y = 3$.

Well done if you follow all the working.

LITERAL EQUATIONS

We shall look ahead to a future section by briefly using the techniques of solving linear equations to solve literal equations. A literal equation is an equation in which variables replace numbers, for example $px + q = r$. We shall solve two literal equations by showing the solutions alongside similar linear equations of types we have already seen.

EXAMPLE:

Solve for x (a) $3x + 5 = 12$ (b) $px + q = r$.

SOLUTION:

(a) $3x + 5 = 12 \Rightarrow 3x = 12 - 5 \Rightarrow 3x = 7 \Rightarrow x = \frac{7}{3}$

(b) $px + q = r \Rightarrow px = r - q \Rightarrow x = \frac{r - q}{p}$.

EXAMPLE:

Solve for x (a) $3(x + 2) = 20$ (b) $p(x + q) = r$.

SOLUTION:

(a) $3(x + 2) = 20 \Rightarrow 3x + 6 = 20 \Rightarrow 3x = 20 - 6 \Rightarrow 3x = 14 \Rightarrow x = \frac{14}{3}$

(b) $p(x + q) = r \Rightarrow px + pq = r \Rightarrow px = r - pq \Rightarrow x = \frac{r - pq}{p}$.

The techniques used in solving literal equations will prove useful when we look at how to change the subject of a formula later.

THINGS TO DO AND THINK ABOUT

Solve algebraically the following equations.

1. (a) $\frac{4}{x} = \frac{2}{7}$ (b) $\frac{3x}{4} = \frac{2}{3}$ (c) $\frac{2}{3}x = \frac{1}{2}$ (d) $\frac{x}{6} = 5$ (e) $\frac{x + 2}{4} = \frac{x + 3}{2}$ (f) $\frac{x}{5} - \frac{1}{2} = \frac{2x}{3}$.

2. Solve for x (a) $a(x - c) = d$ (b) $\frac{x}{7} = \frac{h}{k}$.

3. Solve the equation $\frac{1}{2}(x + 2) = \frac{2}{3}(x + 5) + \frac{1}{6}(x + 1)$.

 ONLINE TEST

Take the 'Further Equations' test online at www. brightredbooks.net/N5Maths

INEQUATIONS

WHAT IS AN INEQUALITY?

In the previous sections, we considered equations and looked at how an equation could be represented by a set of balanced scales. In this section, we shall investigate inequalities. They can be represented by a set of scales which are not balanced, that is, one side of the scales is greater or less than the other.

To identify inequalities, we must use the following symbols:

- $x > 3$ means x is greater than 3
- $x < 3$ means x is less than 3
- $x \geqslant 3$ means x is greater than or equal to 3
- $x \leqslant 3$ means x is less than or equal to 3.

Note also that $x \neq 3$ means x is not equal to 3.

15 Inequalities occur frequently in real–life situations. For example, if you saw this road sign as you drove through a country park, it would indicate that the maximum speed limit for driving in the park is 15 miles per hour.

A mathematician might write this as the inequality $S \leqslant 15$ where S represents speed, indicating that any speed equal to or less than 15 miles per hour is legal.

If we return to the set of scales shown above, they lead to the sentence $3x + 5 > 11$. This is an example of an **inequation**. In this section, we shall study how to solve inequations. The good news is that, for the most part, linear inequations can be solved in exactly the same way as linear equations. In other words, whatever you do to one side of an inequation, that is add, subtract, multiply or divide, you must do the same to the other side to maintain the 'out-of-balance' condition.

EXAMPLE:

Solve the inequation $3x + 5 > 11$.

SOLUTION:

$3x + 5 > 11$

$\Rightarrow 3x > 11 - 5$

$\Rightarrow 3x > 6$

$\Rightarrow x > 2$.

This should look familiar. The working is identical to the example earlier when we solved the equation $3x + 5 = 11$, with one important difference, namely the equal signs have all changed to > signs.

DON'T FORGET

Linear inequations and linear equations can be solved for the most part in the same way.

EXAMPLE:

Solve the inequation $4(3x + 5) \leqslant 3(3x + 7)$.

SOLUTION:

$4(3x + 5) \leqslant 3(3x + 7) \Rightarrow 12x + 20 \leqslant 9x + 21 \Rightarrow 12x - 9x \leqslant 21 - 20$

$\Rightarrow 3x \leqslant 1$

$\Rightarrow x \leqslant \frac{1}{3}$.

DIVIDING AN INEQUALITY BY A NEGATIVE NUMBER

You have been told that you can use the same methods for solving inequations as you did for solving equations. Although this is true, there is an important situation which can arise where you have to be very careful. Consider the true statement $10 > 8$. Suppose we divide both sides of this inequality by -2. Is it true that $-5 > -4$? Obviously it is not true, as $-5 < -4$. It is therefore the case that when you divide throughout an inequality by a negative number, you must reverse the direction of the inequality symbol. The same thing happens when you multiply both sides of an inequality by a negative number. You must use this idea to solve certain inequations.

EXAMPLE:

Solve the inequation $2y - 10 \geqslant 7y$.

SOLUTION:

$$2y - 10 \geqslant 7y$$
$$\Rightarrow 2y - 7y \geqslant 10$$
$$\Rightarrow -5y \geqslant 10$$
$$\div(-5) \quad \div(-5) \qquad \text{(Divide both sides by } -5)$$
$$\Rightarrow \qquad y \leqslant -2. \qquad \text{(Reverse the direction of the inequality symbol)}$$

NOTE: It is worth pointing out that the above inequation could be solved in a different way by collecting the variables on the right side of the inequality symbol.

Let's look at an alternative solution:

$$2y - 10 \geqslant 7y$$
$$\Rightarrow -10 \geqslant 7y - 2y$$
$$\Rightarrow -10 \geqslant 5y$$
$$\div 5 \quad \div 5 \qquad \text{(Divide both sides by 5)}$$
$$\Rightarrow -2 \geqslant y$$
$$\Rightarrow \quad y \leqslant -2.$$

The intention here is not to confuse you but to point out that, as in many other areas of mathematics, there are often alternative ways to arrive at the correct solution. You should focus on finding the way that consistently enables you to arrive at the correct solution.

EXAMPLE:

Solve the inequation $3(a + 4) < 5(a + 2)$.

SOLUTION:

$$3(a + 4) < 5(a + 2) \Rightarrow 3a + 12 < 5a + 10 \Rightarrow 3a - 5a < 10 - 12 \Rightarrow -2a < -2$$
$$\Rightarrow a > 1.$$

Note how we had to divide throughout by -2 to complete the inequation, and this led to the reversal of the direction of the inequality symbol.

THINGS TO DO AND THINK ABOUT

Solve the following inequations.

(a) $5x + 3 > 38$

(b) $7x - 2 > 3x + 18$

(c) $4(x + 2) \leqslant 2(x - 10)$

(d) $3x - 7 \geqslant 4x$

(e) $2x < 12 + 5x$

(f) $4(x + 2) \leqslant 6(x + 3)$.

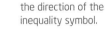

DON'T FORGET

If you have to divide (or multiply) both sides of an inequation by a negative number, you must reverse the direction of the inequality symbol.

VIDEO LINK

Watch 'Solving Inequalities' at www.brightredbooks.net/N5Maths

ONLINE TEST

Take the 'Inequations' test online at www.brightredbooks.net/N5Maths

SIMULTANEOUS EQUATIONS

A GRAPHICAL METHOD

Simultaneous equations are equations involving two or more variables. In this course, we shall only consider simultaneous equations with two variables. We will have to solve such simultaneous equations and find the values of the two unknown quantities. The unknowns will have the same values in both equations. Simultaneous equations are sometimes referred to as a system of equations.

In this section, we will study how to solve simultaneous equations graphically.

EXAMPLE:

Solve graphically the system of equations $2x + y = 12$
$$y = x - 3.$$

SOLUTION:

Both equations are linear equations; that is, they are the equations of straight lines.
We start by finding some points which lie on each line.

Consider the linear equation $2x + y = 12$. We must find the coordinates of some points on this line (think back to drawing lines from an earlier section).

We find that when $y = 0$, $2x = 12 \Rightarrow x = 6$. When $x = 0$, $y = 12$. Hence this line cuts the x- and y-axes at $(6, 0)$ and $(0, 12)$ respectively. Find a third point as a check.

When $x = 3$, $(2 \times 3) + y = 12 \Rightarrow 6 + y = 12 \Rightarrow y = 6$. Hence $(3, 6)$ lies on this line. The results could be put into a table.

x	0	3	6
y	12	6	0

Now concentrate on the linear equation $y = x - 3$. Choose some simple values for x, say 0, 4 and 8, and form a table as shown below.

x	0	4	8
y	-3	1	5

This line passes through the points $(0, -3)$, $(4, 1)$ and $(8, 5)$. Now draw both lines on the same grid and label each line.

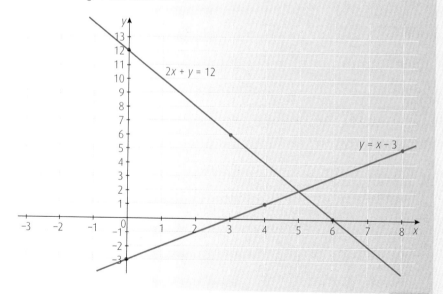

DON'T FORGET

Always check the solutions to simultaneous equations.

contd

By inspection, we can see that the two lines intersect at the point with coordinates (5, 2). Hence the solution of the simultaneous equations is $x = 5$, $y = 2$.

Although this is a lengthy process, with practice you should soon become efficient. It is possible that you might see the solution quicker sometimes if the same point appears in both tables. However, if you are asked to find the solution graphically, you must continue and complete the grid with both lines clearly drawn.

Note that if you are asked to find the point of intersection of the straight lines with equations $2x + y = 12$ and $y = x - 3$, this would be an acceptable way of showing that the solution is (5, 2).

An important aspect of solving simultaneous equations is the fact that you can check if your solutions are correct. If both solutions ($x = 5$, $y = 2$) are correct, then they should fit into the original equations to give true statements.

Check

Substitute ($x = 5$, $y = 2$) into $2x + y = 12 \Rightarrow (2 \times 5) + 2 = 10 + 2 = 12$.

Substitute ($x = 5$, $y = 2$) into $y = x - 3 \Rightarrow 2 = 5 - 3$.

As both statements are true, we can see that the solutions are correct. If the statements are not true, then you must go back and check your prior working.

VIDEO LINK

Watch 'Solving simultaneous equations by graph' at www. brightredbooks.net/N5Maths

SOLVING SIMULTANEOUS EQUATIONS BY SUBSTITUTION

A non-graphical method is illustrated to show an alternative way of arriving at the solution.

SOLUTION:

Substitute $y = x - 3$ into the equation $2x + y = 12$, that is, replace y by $x - 3$ in the equation $2x + y = 12$.

Hence $2x + y = 12 \Rightarrow 2x + x - 3 = 12 \Rightarrow 3x - 3 = 12 \Rightarrow 3x = 12 + 3 \Rightarrow 3x = 15 \Rightarrow x = 5$.

Once we have found that $x = 5$, it is simple to find that $y = 2$ by substitution. Note that this method would not be acceptable if you were asked to solve the equations graphically. The method of substitution is sometimes useful as a check, but only if the equation formed is easy to solve.

ONLINE TEST

Take the 'Simultaneous Equations' test online at www.brightredbooks.net/N5Maths

THINGS TO DO AND THINK ABOUT

1. Solve graphically the simultaneous equations
   $x + y = 9$
   $y = 2x.$

2. Solve graphically the simultaneous equations
   $y = 2x - 1$
   $x = 4.$

3. Solve graphically the simultaneous equations
   $3x + y = 12$
   $x + 3y = 12.$

4. Solve by substitution the simultaneous equations
   $y = 2x + 1$
   $3x + 2y = 37.$

MORE SIMULTANEOUS EQUATIONS

ELIMINATION

The method of solving simultaneous equations graphically is not always suitable, for example when the solutions are not integers. In this case, it may be difficult to read off fractional values on a grid. Similarly, this method would be unsuitable if the solution involved large numbers. The method of substitution can also be tricky if the coefficients of the variables are not 1, as this will lead to fractions appearing in the working. For these reasons, the recommended method for solving simultaneous equations, assuming you are given a choice, is elimination. The method of elimination involves 'getting rid of' one of the variables in order to reduce to a simpler equation in one variable only. If you are asked to solve the equations algebraically, this would be ideal.

In this method, the two equations must be lined up with the variables on the left side and the constants on the right. The equations will be numbered (1) and (2) and then scaled, that is multiplied, if necessary, to make the coefficients of one of the variables equal. The resulting equations can then be added or subtracted to get rid of (or eliminate) one of the variables. After that is done, the task is fairly straightforward. Study the example below.

EXAMPLE:

Solve algebraically the simultaneous equations
$$5x + y = 17$$
$$3x - 2y = 18.$$

SOLUTION:

$$5x + y = 17 \qquad (1)$$
$$3x - 2y = 18 \qquad (2)$$
$$(1) \times 2: \quad 10x + 2y = 34 \qquad (3)$$
$$(2) \times 1: \quad 3x - 2y = 18 \qquad (4)$$

$$(3) + (4): \quad 13x = 52$$
$$\Rightarrow x = \frac{52}{13} = 4.$$

Now substitute $x = 4$ into equation (1)
$$5x + y = 17$$
$$\Rightarrow 5 \times 4 + y = 17$$
$$\Rightarrow 20 + y = 17$$
$$\Rightarrow y = 17 - 20 = -3.$$

Solution is $x = 4$, $y = -3$.

Check

Substitute both values into equation (2): $3x - 2y = 18$
$$\Rightarrow 3 \times 4 - 2 \times (-3) = 12 + 6 = 18.$$

DON'T FORGET

We added equations (3) and (4) because $2y + (-2y) = 0$ and this eliminates the variable y from the equations, leaving us with a linear equation. It is normal after you have evaluated the first variable to substitute in equation (1) and check in equation (2). Here is another example. Study it carefully.

EXAMPLE:

Solve algebraically the simultaneous equations
$$3x - 4y = 8$$
$$2x - 5y = -4.$$

SOLUTION:

$$3x - 4y = 8 \qquad (1)$$
$$2x - 5y = -4 \qquad (2)$$
$$(1) \times 5: \quad 15x - 20y = 40 \qquad (3)$$
$$(2) \times 4: \quad 8x - 20y = -16 \qquad (4)$$

contd

(3) – (4): $\qquad 7x = 56$

$$\Rightarrow x = \frac{56}{7} = 8.$$

Now substitute $x = 8$ into equation (1)

$$3x - 4y = 8$$
$$\Rightarrow 3 \times 8 - 4y = 8$$
$$\Rightarrow 24 - 4y = 8$$
$$\Rightarrow -4y = 8 - 24 = -16$$
$$\Rightarrow y = \frac{-16}{-4} = 4$$

Solution is $x = 8$, $y = 4$.

Check

Substitute both values into equation (2): $2x - 5y = -4$

$$\Rightarrow 2 \times 8 - 5 \times 4 = 16 - 20 = -4.$$

> **BEWARE:** This time, we had to subtract the equations (3) and (4), as $-20y - (-20y) = 0$. Hence y is eliminated. Remember: if the equations are lined up properly and the two signs are the same, subtract the equations. Be particularly careful if you have to subtract a negative quantity, for example $40 - (-16) = 56$.

Finally, we return to equations from the previous section when we solved simultaneous equations graphically. This time, we will use elimination.

EXAMPLE:

Find the point of intersection of the straight lines with equations $2x + y = 12$ and $y = x - 3$.

SOLUTION:

The elimination method is ideal for this type of question. You must rearrange $y = x - 3$ into the form $x - y = 3$ at the start. Be careful when rearranging, as an error here will affect everything to follow.

$$2x + y = 12 \qquad (1)$$
$$x - y = 3 \qquad (2)$$

(1) + (2) $\qquad 3x = 15$

$$\Rightarrow x = \frac{15}{3} = 5.$$

Now substitute $x = 5$ into equation (1)

$$2x + y = 12$$
$$\Rightarrow 2 \times 5 + y = 12$$
$$\Rightarrow 10 + y = 12$$
$$\Rightarrow y = 2$$

Solution is $x = 5$, $y = 2$.

Check

Substitute both values into equation (2): $x - y = 3$

$$\Rightarrow 5 - 2 = 3$$

Hence the point of intersection of the lines is $(5, 2)$.

VIDEO LINK

Watch another example: 'Solving Simultaneous Equations by Elimination Example 1' at www.brightredbooks.net/N5Maths

DON'T FORGET

To use the elimination method, line up the equations properly, scale the equations if necessary, then add if the signs are different, subtract if the signs are the same. If you have been taught to multiply one equation by a negative number and add in the latter case, that is fine. Remember to check your solutions. If the answers turn out to have unusual fractions, you have probably made a mistake, so check your working carefully.

ONLINE TEST

Take the 'More Simultaneous Equations' test online at www.brightredbooks.net/N5Maths

THINGS TO DO AND THINK ABOUT

Solve the following simultaneous equations.

(a) $2x + y = 9$
$\quad 6x - 5y = 35$

(b) $4x + 3y = 13$
$\quad 3x + 2y = 9$.

PROBLEMS INVOLVING SIMULTANEOUS EQUATIONS

REAL-LIFE SITUATIONS

Many problems which occur in real life can be solved using simultaneous equations. You will have to make up the equations first, however.

EXAMPLE:

At the cinema, the Pollock family buys 3 adult tickets and 5 child tickets. The total cost is £51·10.

Let x pounds be the cost of an adult ticket and y pounds be the cost of a child ticket.

(a) Write down an equation in x and y which satisfies this condition.

The Ahmed family buys 4 adult tickets and 3 child tickets. The total cost is £48·70.

(b) Write down a second equation in x and y which satisfies this condition.

(c) Find the cost of an adult ticket and the cost of a child ticket.

SOLUTION:

(a) $\qquad 3x + 5y = 51\cdot10$

(b) $\qquad 4x + 3y = 48\cdot70$

(c) $\qquad 3x + 5y = 51\cdot10 \qquad$ (1)

$\qquad\qquad 4x + 3y = 48\cdot70 \qquad$ (2)

$(1) \times 3: \qquad 9x + 15y = 153\cdot30 \qquad$ (3)

$(2) \times 5: \qquad 20x + 15y = 243\cdot50 \qquad$ (4)

$(4) - (3): \qquad 11x = 90\cdot20$

$$\Rightarrow x = \frac{90\cdot20}{11} = 8\cdot20.$$

Now substitute $x = 8\cdot20$ into equation (1)

$$3x + 5y = 51\cdot10$$

$$\Rightarrow 3 \times 8\cdot20 + 5y = 51\cdot10$$

$$\Rightarrow 24\cdot60 + 5y = 51\cdot10$$

$$\Rightarrow 5y = 51\cdot10 - 24\cdot60 = 26\cdot50$$

$$\Rightarrow y = \frac{26\cdot50}{5} = 5\cdot30.$$

Solution is $x = 8\cdot20$, $y = 5\cdot30$.

Hence an adult ticket costs £8·20 and a child ticket costs £5·30.

Check

Substitute both values into equation (2):

$4x + 3y = 48\cdot70 \Rightarrow 4 \times 8\cdot20 + 3 \times 5\cdot30 = 32\cdot80 + 15\cdot90 = 48\cdot70.$

DON'T FORGET

On completing the working for this type of problem, it is important that you communicate the solution in words.

contd

EXAMPLE:

Joan has been saving 20p coins and 10p coins. She has a total of 120 coins so far. The total value of all these coins is £19·10.

Find the number of 20p coins and the number of 10p coins Joan has.

SOLUTION:

Simultaneous equations can be used to solve this type of problem with two unknowns.

Let x be the number of 20p coins and y be the number of 10p coins. Now you can form two equations.

$$x + y = 120 \qquad (1)$$
$$20x + 10y = 1910 \qquad (2)$$

(1) × 10: $10x + 10y = 1200 \qquad (3)$

(2) − (3): $10x = 710$

$$\Rightarrow x = \frac{710}{10} = 71.$$

Now substitute $x = 71$ into equation (1)

$$x + y = 120$$
$$\Rightarrow 71 + y = 120$$
$$\Rightarrow y = 120 - 71$$
$$\Rightarrow y = 49.$$

Solution is $x = 71$, $y = 49$.

The number of 20p coins Joan has is 71 and the number of 10p coins is 49.

Check

Substitute both values into equation (2): $20x + 10y = 1910$

$$\Rightarrow 20 \times 71 + 10 \times 49 = 1420 + 490 = 1910.$$

> **BEWARE:** This type of question can lead to some common errors. Many students would write equation (2) as $20x + 10y = 19·10$. This would lead to a negative answer. Therefore be careful when forming equations in which pounds and pence are involved, and make sure that the units are consistent. Other students would write equation (1) in error as $20x + 10y = 120$ and equation (2) correctly as $20x + 10y = 1910$. These two equations cannot be solved, as they represent parallel lines and do not intersect. So watch out for this type of error. With practice, solving simultaneous equations should become straightforward.

VIDEO LINK

Watch 'Another money problem' at www. brightredbooks.net/N5Maths

ONLINE TEST

Take the 'Problems Involving Simultaneous Equations' test online at www. brightredbooks.net/N5Maths

THINGS TO DO AND THINK ABOUT

A theatre has 500 seats which are either in the stalls or the circle.

Let x be the number of seats in the stalls and y be the number of seats in the circle.

(a) Write down an equation in x and y which satisfies this condition.

A seat in the stalls costs £18 and a seat in the circle costs £25.

When all the seats are sold, the ticket sales are £10 120.

(b) Write down a second equation in x and y which satisfies this condition.

(c) Find the cost of a ticket for the stalls and the cost of a ticket for the circle.

CHANGING THE SUBJECT OF A FORMULA

DON'T FORGET

The correct order of operations is essential when carrying out calculations and using formulae. A useful memory aid is the mnemonic BODMAS:

B	Brackets first
O	Orders (Powers, Roots)
DM	Division and Multiplication
AS	Addition and Subtraction

VIDEO LINK

Listen to the 'BODMAS Song' at www.brightredbooks.net/N5Maths

USING A FORMULA

We have already met and used several formulae in this guide, such as formulae involving circles and volume as well as the gradient formula. You should be able to use formulae by substituting given values into a formula and carrying out calculations. Consider the example below, which looks at one of the laws of motion used in physics.

EXAMPLE:

The distance, s metres, travelled by an accelerating object is given by the formula
$$s = ut + \tfrac{1}{2}at^2$$
where u metres per second is the initial velocity, t seconds is the time taken and a metres per second per second is the acceleration.

Calculate s when $u = 6$, $t = 10$ and $a = 8$.

Do not use a calculator.

SOLUTION:

$s = ut + \tfrac{1}{2}at^2 = 6 \times 10 + \tfrac{1}{2} \times 8 \times 10^2 = 6 \times 10 + \tfrac{1}{2} \times 8 \times 100 = 60 + 400 = 460$

CHANGING THE SUBJECT OF A FORMULA

In the formula from the first example above, s was the subject of the formula. It is often useful to be able to change the subject of a formula to one of the other variables. For example, earlier we studied volume and did the following example on working back.

The volume of a cone is 1500 cubic centimetres. Its height is 8 centimetres. Calculate its radius.

This example could be done by changing the subject of the formula for the volume of a cone, $V = \tfrac{1}{3}\pi r^2 h$, to r.

EXAMPLE:

Change the subject of the formula $V = \tfrac{1}{3}\pi r^2 h$ to r.

SOLUTION:

Think of $V = \tfrac{1}{3}\pi r^2 h$ as $\tfrac{V}{1} = \tfrac{\pi r^2 h}{3}$, then cross–multiply.

This leads to $\tfrac{V}{1} = \tfrac{\pi r^2 h}{3} \Rightarrow 3V = \pi r^2 h \Rightarrow \pi r^2 h = 3V \Rightarrow r^2 = \tfrac{3V}{\pi h} \Rightarrow r = \sqrt{\tfrac{3V}{\pi h}}$.

NOTE: The earlier example can then be done by substituting $V = 1500$ and $h = 8$ into this new formula. Check that this leads to $r = 13\cdot38$, the same as before.

HOW TO CHANGE THE SUBJECT OF A FORMULA

If you consider the above example, you will see that it was solved like a literal equation. Remember that a literal equation is one in which letters replace numbers. We were then able to use cross–multiplication, as we had two equal fractions. We can think of changing the subject of a formula as solving a literal equation.

EXAMPLE:

The perimeter of a rectangle can be found using the formula $P = 2(l + b)$. Change the subject of this formula to l.

contd

SOLUTION:

$$P = 2(l + b)$$
$$\Rightarrow P = 2l + 2b \quad \text{(multiply out the brackets)}$$
$$\Rightarrow 2l + 2b = P \quad \text{(swap the sides)}$$
$$\Rightarrow 2l = P - 2b \quad \text{(subtract } 2b \text{ from both sides)}$$
$$\Rightarrow l = \frac{P - 2b}{2} \quad \text{(divide both sides by 2).}$$

EXAMPLE:

Change the subject of the formula $p = 6q^2 + r$ to q.

SOLUTION:

$$p = 6q^2 + r$$
$$\Rightarrow 6q^2 + r = p \quad \text{(swap the sides)}$$
$$\Rightarrow 6q^2 = p - r \quad \text{(subtract } r \text{ from both sides)}$$
$$\Rightarrow q^2 = \frac{p - r}{6} \quad \text{(divide both sides by 6)}$$
$$\Rightarrow q = \sqrt{\frac{p - r}{6}} \quad \text{(take the square root of both sides)}$$

EXAMPLE:

Change the subject of the formula $A = \frac{1}{2}bh$ to h.

SOLUTION:

$$A = \frac{1}{2}bh$$
$$\Rightarrow 2A = bh \quad \text{(cross–multiply)}$$
$$\Rightarrow h = \frac{2A}{b} \quad \text{(swap the sides and divide by } b\text{).}$$

> **DON'T FORGET**
>
> Changing the subject of a formula is like solving an equation. Whatever operation is carried out on one side, you must do the same to the other side.

Other methods of changing the subject are often taught, including a doing–and–undoing approach. As always, use the method that you feel most comfortable with.

EXAMPLE:

The volume of a cylinder is 5000 cubic centimetres. Its radius is 15 centimetres. Calculate its height.

SOLUTION:

This time, we will change the subject of the formula $V = \pi r^2 h$ to h.
$$V = \pi r^2 h \Rightarrow \pi r^2 h = V \Rightarrow h = \frac{V}{\pi r^2}$$
Hence $h = \frac{V}{\pi r^2} \Rightarrow \frac{5000}{\pi \times 15^2} = \frac{5000}{706 \cdot 858} = 7 \cdot 073\,6$
Therefore the height of the cylinder is $7 \cdot 1$ cm (correct to 2 sig. figs).

> **ONLINE TEST**
>
> Take the 'Changing the Subject of a Formula' test online at www.brightredbooks.net/N5Maths

THINGS TO DO AND THINK ABOUT

Change the subject in each of the following formulae.

(a) $d = ab + c$ to a (b) $m(r - s) = t$ to r (c) $A = 4\pi r^2$ to r

(d) $C = \pi d$ to d (e) $V = lbh$ to b (f) $a = bc^2 + d$ to c

(g) $e = mc^2$ to c (h) $A = \frac{kp^2}{2}$ to p (i) $V = \frac{1}{3}Ah$ to A

(j) $y = \frac{1}{2}(2x + z)$ to x.

QUADRATIC FUNCTIONS

THE EQUATION OF A QUADRATIC FUNCTION

We briefly looked at quadratic functions earlier in the sections on factorising trinomials, completing the square and functional notation. Now we shall study such functions in greater detail. The general equation of a quadratic function is

$$f(x) = ax^2 + bx + c,$$

where a cannot equal zero. Remember also that $y = f(x)$ from the section on functional notation.

Hence functions such as $f(x) = x^2 + 3x + 4$, $f(x) = 2x^2 - 5x$, $f(x) = 3x^2 + 12$, $f(x) = 5x^2$ and $f(x) = 6 + 4x - x^2$ are all examples of quadratic functions. We shall consider a very straightforward quadratic function, that is $f(x) = x^2$. We can draw the graph of this function by creating a table of values.

x	-3	-2	-1	0	1	2	3
$f(x) = x^2$	9	4	1	0	1	4	9

By plotting points from this table, we can see the graph of the function $f(x) = x^2$.

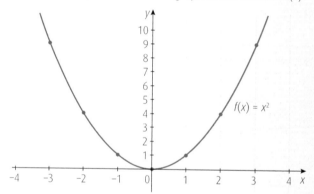

As mentioned earlier, this type of graph is called a **parabola**. The graphs of all quadratic functions are parabolas. A parabola is a smooth symmetrical curve with **an axis of symmetry**. For the above parabola, the axis of symmetry is the y-axis, that is the straight line with equation $x = 0$. All parabolas have turning points. The above parabola has a **minimum turning point** with coordinates $(0, 0)$. Parabolas with minimum turning points are sometimes described as $\cup$-shaped. The graph of the function $f(x) = ax^2 + bx + c$ will be $\cup$-shaped if $a > 0$.

If the quadratic function $f(x) = ax^2 + bx + c$ is such that $a < 0$, for example $f(x) = -x^2 + 5$, then the parabola will be $\cap$-shaped and have a maximum turning point.

EXAMPLE:

Part of the graph of $y = ax^2$ is shown in the diagram.

If the point $(2, 8)$ lies on the graph, find the value of a.

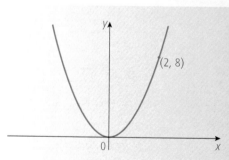

SOLUTION:

Substitute $x = 2$, $y = 8$ into the equation $y = ax^2$.

Hence $y = ax^2 \Rightarrow 8 = a \times 2^2 \Rightarrow 8 = 4a \Rightarrow a = 2$.

EQUATIONS OF THE FORM $y = (x + p)^2 + q$

VIDEO LINK

For an example of this, noting that the term 'vertex' is used for turning point, watch 'Putting a quadratic function in standard form' at www.brightredbooks.net/N5Maths

When we studied the topic of completing the square, we looked at the example where we found that $x^2 - 12x + 25 = (x - 6)^2 - 11$. It would be useful to remind yourself of this section by looking at page 31. A sketch of the graph of $y = x^2 - 12x + 25$ had a minimum turning point at $(6, -11)$. We were able to tell this by considering the same equation in the form $y = (x - 6)^2 - 11$. By completing the square, we can find out the coordinates of the turning point of a parabola.

EXAMPLE:

Express the equation $y = x^2 + 8x - 4$ in the form $y = (x + p)^2 + q$. Hence state the coordinates of the turning point on the graph of the quadratic function with equation $y = x^2 + 8x - 4$.

SOLUTION:

Use the technique of completing the square.

$y = x^2 + 8x - 4 \Rightarrow y = x^2 + 8x + 16 - 16 - 4 \Rightarrow y = (x + 4)^2 - 20$.

As this has a minimum value of -20 when $x = -4$, the coordinates of the turning point are $(-4, -20)$.

NOTE: We know that the turning point of the graph is a minimum turning point by inspecting the equation $y = x^2 + 8x - 4$. Remember that the graph of the function $f(x) = ax^2 + bx + c$ will be $\cup$-shaped if $a > 0$. In the case of $y = x^2 + 8x - 4$, the coefficient of the x^2 term is greater than zero.

DON'T FORGET

The graph of the quadratic function $f(x) = a(x + p)^2 + q$ has a turning point at $(-p, q)$. It will be a minimum turning point if the coefficient of the x^2 term is greater than zero and a maximum turning point if the coefficient of the x^2 term is less than zero. The axis of symmetry passes through the turning point and has equation $x = -p$.

EXAMPLE:

The diagram shows part of the graph of $y = 15 - (x - 4)^2$.

(a) State the coordinates of the maximum turning point.

(b) State the equation of the axis of symmetry of the graph.

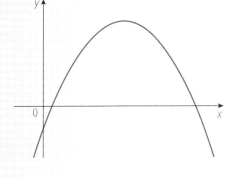

SOLUTION:

(a) $(4, 15)$ (b) $x = 4$.

ONLINE TEST

Take the 'Quadratic Functions' test online at www.brightredbooks.net/N5Maths

 THINGS TO DO AND THINK ABOUT

1. By completing the table below, sketch the graph of the quadratic function $y = -x^2$.

x	-3	-2	-1	0	1	2	3
$y = -x^2$							

2. The point $(3, -45)$ lies on the graph of the function $y = kx^2$. Find the value of k.

3. Express the equation $y = x^2 + 2x + 5$ in the form $y = (x + p)^2 + q$. Hence state the coordinates of the turning point on the graph of the quadratic function with equation $y = x^2 + 2x + 5$.

4. A quadratic function has equation $f(x) = (x + 2)^2 + 7$.

   (a) State the coordinates of the minimum turning point of the graph of this function.

   (b) State the equation of the axis of symmetry of the graph.

MORE ABOUT QUADRATIC FUNCTIONS

SOME IMPORTANT FACTS

- The general equation of a quadratic function is $y = ax^2 + bx + c$ or $f(x) = ax^2 + bx + c$
- By completing the square, the equation can be written as $y = a(x + p)^2 + q$
- The graph of a quadratic function is a parabola
- The parabola with equation $y = a(x + p)^2 + q$ has a turning point with coordinates $(-p, q)$
- The axis of symmetry of the parabola with equation $y = a(x + p)^2 + q$ is $x = -p$
- If the coefficient of the x^2 term in a quadratic function is positive, the graph has a minimum turning point
- If the coefficient of the x^2 term in a quadratic function is negative, the graph has a maximum turning point

DRAWING THE GRAPH OF A QUADRATIC FUNCTION

The graph of a quadratic function may be drawn by using a graphical calculator. Graphics software can be used to investigate graphs and zoom in on particular points on the graph. However, at present we shall concentrate on sketching such graphs without a calculator. In the section on functional notation on page 53, we drew the graph of the quadratic function $f(x) = x^2 - 12x + 25$ by constructing a table of values. This method can be used to get a fairly accurate drawing of the graph of a quadratic function.

EXAMPLE:

Draw the graph of the quadratic function $y = x^2 - 2x - 8$ by choosing values of x from −3 to 5.

SOLUTION:

x	−3	−2	−1	0	1	2	3	4	5
x^2	9	4	1	0	1	4	9	16	25
$-2x$	6	4	2	0	− 2	−4	−6	−8	−10
-8	−8	−8	−8	−8	− 8	−8	−8	−8	−8
$y = x^2 - 2x - 8$	7	0	−5	−8	−9	−8	−5	0	7

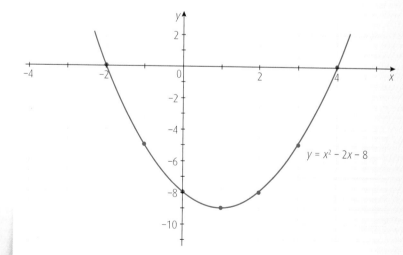

$y = x^2 - 2x - 8$

This method of constructing a detailed table of values and plotting points gives an accurate picture of the graph. You can also check the symmetry of the y-values as you work through the table. However, this method is time-consuming (it is unlikely to be tested in an exam situation), and usually a simpler sketch will be adequate.

To get a quick sketch, we need the turning point (or vertex) of the parabola. If the equation is given in the form $y = (x + p)^2 + q$, this will be easy. Otherwise we must complete the square. For the above quadratic function, we find the turning point as follows:

contd

$y = x^2 - 2x - 8 = x^2 - 2x + 1 - 1 - 8 = (x - 1)^2 - 9$.

This leads to the turning point at $(1, -9)$. Then decide if the parabola has a maximum or minimum turning point by inspecting the coefficient of the x^2 term. As it is positive (1), the graph has a minimum turning point and is therefore ∪-shaped. To complete a quick sketch, find the coordinates of the point where the graph cuts the y-axis. This can be done very easily by replacing x by zero in the equation $y = x^2 - 2x - 8$, leading to $y = -8$. So, the graph cuts the y-axis at $(0, -8)$. Now that we have the turning point, the shape and the y-intercept, a quick sketch can be drawn.

EXAMPLE:

Sketch the graph of the function $y = x^2 + 8x + 10$.

SOLUTION:

Complete the square: $y = x^2 + 8x + 10 \Rightarrow y = x^2 + 8x + 16 - 16 + 10$

$\Rightarrow y = (x + 4)^2 - 6$.

Hence the turning point has coordinates $(-4, -6)$ and is a minimum turning point, as the coefficient of the x^2 term is positive. By replacing x by zero in the equation $y = x^2 + 8x + 10$ leading to $y = 10$, we find that the graph cuts the y-axis at $(0, 10)$. Now do a sketch.

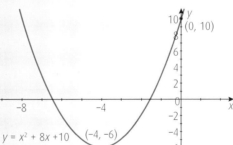

THE ROOTS OF A QUADRATIC EQUATION

The roots of a quadratic equation tell us where the parabola cuts the x-axis. As a graph cuts the x-axis where $y = 0$, they are found by solving the equation $f(x) = 0$. In the example opposite, we looked at the quadratic function $y = x^2 - 2x - 8$. The roots of this function are found by solving the equation $x^2 - 2x - 8 = 0$. This is called a quadratic equation, and the solutions are called the roots of the equation. We shall study how to solve equations like this shortly, but you can see from the table and graph that there are two roots at -2 and 4. When x is replaced by these values in $y = x^2 - 2x - 8$, then $y = 0$.

BEWARE: The graphs of some parabolas do not cross the x-axis. In that case, there are no roots. The graphs of some parabolas touch the x-axis at one point only. In that case, there are equal roots (essentially one root). See the diagrams alongside for examples.

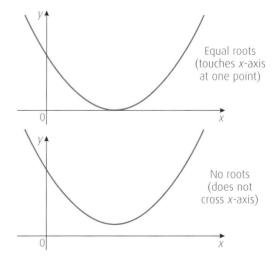

Equal roots (touches x-axis at one point)

No roots (does not cross x-axis)

THINGS TO DO AND THINK ABOUT

1. Draw the graph of the quadratic function $y = x^2 - 6x + 8$ by choosing values of x from 0 to 6.

2. State the roots of the equation $x^2 - 6x + 8 = 0$.

3. Sketch the graph of the function $y = x^2 - 2x - 15$.

SOLVING QUADRATIC EQUATIONS BY FACTORISATION

FINDING THE ROOTS OF A QUADRATIC EQUATION

In this section, we shall study how to solve quadratic equations by factorisation. The solutions of quadratic equations are called the roots of the equation.

These equations cannot be solved in the same way as linear equations. In fact, they are solved on the basis that the solution of the equation $pq = 0$ is either $p = 0$ or $q = 0$.

The type of factorisation could be a common factor, a difference of two squares or a trinomial.

EXAMPLE:

Find the roots of the equation $x^2 + 7x = 0$.

SOLUTION:

Using a common factor, $x^2 + 7x = 0 \Rightarrow x(x + 7) = 0 \Rightarrow$ either $x = 0$ or $x + 7 = 0$
$$\Rightarrow x = 0 \text{ or } x = -7.$$

EXAMPLE:

Find the roots of the equation $9 - x^2 = 0$.

SOLUTION:

$9 - x^2 = 0 \Rightarrow 3^2 - x^2 = 0 \Rightarrow (3 + x)(3 - x) = 0 \Rightarrow$ either $3 + x = 0$ or $3 - x = 0$
$$\Rightarrow x = -3 \text{ or } x = 3.$$

EXAMPLE:

Find the roots of the equation $x^2 - 6x = 7$.

SOLUTION:

The right-hand side must equal zero before factorising, so rearrange as shown:
$x^2 - 6x - 7 = 0 \Rightarrow (x + 1)(x - 7) = 0 \Rightarrow$ either $x + 1 = 0$ or $x - 7 = 0 \Rightarrow x = -1$ or $x = 7$.

EXAMPLE:

Find the coordinates of the points where the graph of the quadratic function $y = 2x^2 + 7x - 15$ cuts the x-axis.

SOLUTION:

The graph cuts the x-axis where $y = 0$.

Hence $2x^2 + 7x - 15 = 0 \Rightarrow (2x - 3)(x + 5) = 0 \Rightarrow 2x - 3 = 0$ or $x + 5 = 0$
$$\Rightarrow x = \tfrac{3}{2} \text{ or } x = -5.$$

The graph cuts the x-axis at the points $(\tfrac{3}{2}, 0)$ and $(-5, 0)$.

Obviously you are going to find this difficult if you cannot factorise a trinomial. If you are in this position, check the earlier sections on factorisation, and practise some extra examples.

> **DON'T FORGET**
>
> The roots of the quadratic equation $(x - d)(x - e) = 0$ are $x = d$ and $x = e$. These roots give you the points where the graph cuts the x-axis.

> **DON'T FORGET**
>
> If the solution to a real-life problem has an impossible solution, you must mention it and explain that you are rejecting it. Also, it is important that you make the right-hand side of a quadratic equation equal to zero before factorising.

MORE GRAPH-SKETCHING

When we have a quadratic function which can be factorised, there is another way of drawing a quick sketch of the graph of the function using the roots and symmetry.

contd

EXAMPLE:

A quadratic function is given by $y = x^2 - 12x + 20$.
Draw a sketch of the graph of this function.

SOLUTION:

The graph cuts the x-axis when $y = 0$.

Hence $x^2 - 12x + 20 = 0 \Rightarrow (x - 2)(x - 10) = 0 \Rightarrow x = 2$ or 10.

The equation of the axis of symmetry is $x = 6$ (midway between 2 and 10).

Substitute $x = 6$ into $y = x^2 - 12x + 20 \Rightarrow y = 6^2 - (12 \times 6) + 20$
$$= 36 - 72 + 20 = -16.$$

Hence the coordinates of the turning point are (6, –16). It is a minimum turning point because the coefficient of the x^2 term is positive. The graph also cuts the y-axis when $x = 0$, leading to a y-intercept of 20. The graph is shown below.

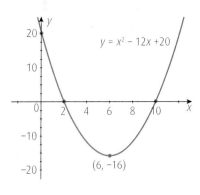

SOLVING PROBLEMS USING QUADRATIC EQUATIONS

Many real-life problems can be solved using quadratic equations.

EXAMPLE:

A rocket is fired from a clifftop into the sea. The diagram shows its path after it is fired. The height, h metres, of the rocket above sea level is given by $h(t) = 64 + 12t - t^2$ where t seconds is the time after firing.
Calculate algebraically the time taken for the rocket to enter the sea.

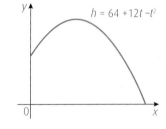

SOLUTION:

The rocket will enter the sea when $h = 0$, so solve the equation $64 + 12t - t^2 = 0$.

$64 + 12t - t^2 = 0 \Rightarrow (16 - t)(4 + t) = 0 \Rightarrow t = 16$ or -4.

As $t = -4$ is impossible, $t = 16$. Therefore it takes 16 seconds for the rocket to enter the sea.

EXAMPLE:

The length of a rectangle is $(x + 2)$ centimetres. Its breadth is $(x + 8)$ centimetres. If the area of the rectangle is 55 square centimetres, calculate x.

SOLUTION:

As $A = lb$ for a rectangle, solve the equation $(x + 2)(x + 8) = 55$.

$(x + 2)(x + 8) = 55 \Rightarrow x(x + 8) + 2(x + 8) = 55 \Rightarrow x^2 + 8x + 2x + 16 - 55 = 0$

Hence $x^2 + 10x - 39 = 0 \Rightarrow (x - 3)(x + 13) = 0 \Rightarrow x = 3$ or $x = -13$.

As $x = -13$ is impossible, $x = 3$.

NOTE: Solving an equation of this type by trial and error is not acceptable, even if you arrive at the correct solution.

VIDEO LINK

Watch 'Solving quadratic equations by factoring' for two good examples on this at www.brightredbooks.net/N5Maths

ONLINE TEST

Take the 'Solving Quadratic Equations by Factorisation' test online at www.brightredbooks.net/N5Maths

THINGS TO DO AND THINK ABOUT

1. Find the roots of the following quadratic equations:

   (a) $5x^2 - 20x = 0$ (b) $x^2 - 36 = 0$ (c) $36 - 5x - x^2 = 0$ (d) $2x^2 + 7x + 5 = 0$.

2. The base of a triangle is $(x + 4)$ centimetres. Its height is $(x + 6)$ centimetres. If the area of the triangle is 24 square centimetres, calculate x.

SOLVING QUADRATIC FUNCTIONS USING THE FORMULA

SOLVING A QUADRATIC EQUATION BY COMPLETING THE SQUARE

When it is impossible to factorise, we need another method to solve quadratic equations. In such cases, the roots of the equation will be irrational. The following method of finding the roots involves completing the square.

EXAMPLE:

Find the roots of the equation $x^2 + 6x + 2 = 0$.

SOLUTION:

$x^2 + 6x + 2 = 0$

$\Rightarrow x^2 + 6x = -2$ (subtract 2 from each side)

$\Rightarrow x^2 + 6x + 9 = -2 + 9$ (add 9 to each side to complete the square)

$\Rightarrow (x + 3)^2 = 7$ (factorise the perfect square and simplify)

$\Rightarrow x + 3 = \pm\sqrt{7}$ (take the square root of both sides)

$\Rightarrow x = -3 \pm \sqrt{7}$ (subtract 3 from each side).

Hence the roots of the equation are $-3 + \sqrt{7}$ or $-3 - \sqrt{7}$. Note that both roots are irrational numbers. Note too that we must take the positive and negative values of the square root.

While this is a neat and acceptable way of solving quadratic equations, it is not the most common way. If the above process of completing the square is applied to the general quadratic equation $ax^2 + bx + c = 0$, we arrive at a formula which can be used to solve all quadratic equations. The formula, known as **the quadratic formula**, is given below.

The roots of $ax^2 + bx + c = 0$ are $x = \dfrac{-b \pm \sqrt{(b^2 - 4ac)}}{2a}$

DON'T FORGET

If you are asked to solve a quadratic equation giving the roots to 1 or 2 decimal places or some similar rounded solution, this is a hint that the answers are likely to be irrational numbers and that you should use the quadratic formula.

EXAMPLE:

Solve the equation $x^2 + 5x + 3 = 0$, giving the roots correct to one decimal place.

SOLUTION:

$a = 1, b = 5, c = 3$

$x = \dfrac{-b \pm \sqrt{(b^2 - 4ac)}}{2a} \Rightarrow x = \dfrac{-5 \pm \sqrt{(5^2 - 4 \times 1 \times 3)}}{2 \times 1}$

$\Rightarrow x = \dfrac{-5 \pm \sqrt{(25 - 12)}}{2}$

$\Rightarrow x = \dfrac{-5 \pm \sqrt{13}}{2}$

$\Rightarrow x = \dfrac{-5 + \sqrt{13}}{2}$ or $x = \dfrac{-5 - \sqrt{13}}{2}$

$\Rightarrow x = \dfrac{-1 \cdot 394}{2}$ or $x = \dfrac{-8 \cdot 606}{2}$

$\Rightarrow x = -0 \cdot 7$ or $x = -4 \cdot 3$ (correct to 1 decimal place).

EXAMPLE:

Solve the equation $3x^2 - 7x - 4 = 0$, giving the roots correct to 2 decimal places.

SOLUTION:

$a = 3, b = -7, c = -4$

contd

$$x = \frac{-b \pm \sqrt{(b^2 - 4ac)}}{2a} \Rightarrow x = \frac{-(-7) \pm \sqrt{((-7)^2 - 4 \times 3 \times (-4))}}{2 \times 3}$$

$$\Rightarrow x = \frac{7 \pm \sqrt{(49 + 48)}}{6}$$

$$\Rightarrow x = \frac{7 \pm \sqrt{97}}{6}$$

$$\Rightarrow x = \frac{7 + \sqrt{97}}{6} \text{ or } x = \frac{7 - \sqrt{97}}{6}$$

$$\Rightarrow x = \frac{16 \cdot 8489}{6} \text{ or } x = \frac{-2 \cdot 8489}{6}$$

$$\Rightarrow x = 2 \cdot 81 \text{ or } x = -0 \cdot 47 \text{ (correct to 2 decimal places).}$$

This example is slightly more difficult due to the fact that b and c are negative. Read through all the working carefully, checking at each step. It is recommended that, if you have to substitute a negative value into the quadratic formula, you put brackets around it. For example, in the above example, b^2 should be written as $(-7)^2$ and not as -7^2, which has a different value.

SOME COMMON ERRORS

It is unlikely that you will be asked to memorise the quadratic formula, although you probably will if you use it often enough. In an exam situation, it will appear on the list of formulae. However, be careful when copying down the formula. It is quite common for students to make a copying error when writing it down. Two common errors are shown below.

$x = -b \pm \frac{\sqrt{(b^2 - 4ac)}}{2a}$ and $x = -b \pm \sqrt{\frac{(b^2 - 4ac)}{2a}}$ both appear from time to time – and they are wrong. Make sure that the line separating the numerator and denominator goes all the way across.

Often, students go wrong towards the end of the process. Consider $\frac{7 + \sqrt{97}}{6}$ from above. If you key in $7 + \sqrt{97} \div 6$ into your calculator, you will get the wrong solution, as you are only dividing $\sqrt{97}$ by 6 and not $7 + \sqrt{97}$ by 6. You can either key in $(7 + \sqrt{97}) \div 6$ or work out the numerator separately as shown in the above examples.

Remember to round your solutions, but make sure that you do not round too early. Some students in the above example would round $\sqrt{97}$ to 2 decimal places – but any rounding should be kept until the end, or else inaccuracies can occur.

THE QUADRATIC FORMULA AND THE AXIS OF SYMMETRY

We can find a useful fact using the formula by considering the two roots of the equation $ax^2 + bx + c = 0$. The roots are $x = \frac{-b + \sqrt{(b^2 - 4ac)}}{2a}$ and $x = \frac{-b - \sqrt{(b^2 - 4ac)}}{2a}$. We know that the roots tell us where the graph of the function $y = ax^2 + bx + c$ cuts the x-axis. Therefore the axis of symmetry of the graph must lie midway between these roots. By finding the mean of the two roots, we can say that the equation of the axis of symmetry of the quadratic function $y = ax^2 + bx + c$ is $x = -\frac{b}{2a}$.

EXAMPLE:

Find the equation of the axis of symmetry of the graph of $y = x^2 - 6x + 4$.

SOLUTION:

$a = 1, b = -6, c = 4$

The equation of the axis of symmetry is $x = -\frac{b}{2a} \Rightarrow x = -\frac{(-6)}{2 \times 1} \Rightarrow x = 3$.

VIDEO LINK

For a couple of good examples, watch 'Using the Quadratic Formula' www.brightredbooks.net/N5Maths

 VIDEO LINK

For an interesting method of drawing parabolas (the step pattern), watch 'How to graph parabolas' at www.brightredbooks.net/N5Maths

 ONLINE TEST

Take the 'Solving Quadratic Functions using the Formula' test online at www.brightredbooks.net/N5Maths

 ## THINGS TO DO AND THINK ABOUT

Solve the following equations, giving the roots correct to 1 decimal place.

(a) $x^2 + x - 5 = 0$ (b) $x^2 + 10x + 6 = 0$ (c) $4x^2 - 7x + 1 = 0$ (d) $2x^2 - x - 9 = 0$.

THE DISCRIMINANT

WHAT IS THE DISCRIMINANT?

Following on from our look at the quadratic formula in the previous section, we now look at one special part of the formula, namely the part under the square-root sign, $b^2 - 4ac$. This part of the formula is called the **discriminant**, and by studying it we can find out what type of roots there are in a quadratic equation. Mathematicians talk about the nature of the roots.

Remember that the quadratic formula for solving the equation $ax^2 + bx + c = 0$ is given by

$$x = \frac{-b \pm \sqrt{(b^2 - 4ac)}}{2a}.$$

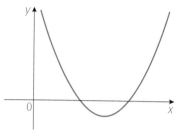

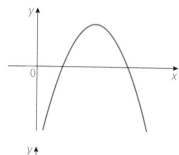

We shall concentrate on the discriminant $b^2 - 4ac$. Suppose $b^2 - 4ac > 0$. In the formula, we find the square root of the discriminant. As we can find the square root of positive numbers and there will be two solutions (one with the + sign and one with the – sign), then two distinct roots will exist. The graph of $y = ax^2 + bx + c$ will cut the x-axis at two points as shown.

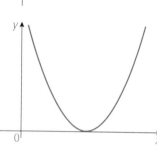

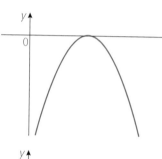

Next, we consider the case where $b^2 - 4ac = 0$. If we replace $b^2 - 4ac$ by 0 in the quadratic formula, then we find that $x = -\frac{b}{2a}$. In this case, the two roots are equal, or there is only one root which is repeated. The graph of $y = ax^2 + bx + c$ will touch the x-axis at one point only, as shown.

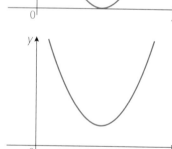

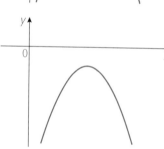

Finally, we consider the case where $b^2 - 4ac < 0$. As we cannot find the square root of negative numbers, there will be no solutions. We say that the quadratic equation has no real roots. The graph of $y = ax^2 + bx + c$ will not cut the x-axis, as shown.

SUMMARY OF RESULTS

For the equation $ax^2 + bx + c = 0$, the discriminant is $b^2 - 4ac$, and

- $b^2 - 4ac > 0 \Rightarrow$ two distinct real roots
- $b^2 - 4ac = 0 \Rightarrow$ equal real roots
- $b^2 - 4ac < 0 \Rightarrow$ no real roots.

DON'T FORGET

The nature of the roots of a quadratic equation depends on whether the discriminant is positive, negative or zero.

When $b^2 - 4ac > 0$ and is also a perfect square, then the roots of the quadratic equation will be rational numbers. This is because we can calculate an exact square root. The expression $ax^2 + bx + c$ will factorise in this case.

When $b^2 - 4ac > 0$ and is not a perfect square, then the roots of the quadratic equation will be irrational numbers.

Note that the condition for real roots is $b^2 - 4ac \geqslant 0$.

PROBLEMS INVOLVING THE DISCRIMINANT

Using the information in the summary, we can quickly find the nature of the roots of a quadratic equation. Problems may involve solving equations or inequations.

EXAMPLE:

What is the nature of the roots of the equation $x^2 + 5x - 2 = 0$?

SOLUTION:

Find the value of $b^2 - 4ac$ using $a = 1$, $b = 5$, $c = -2$.

Hence $b^2 - 4ac = 5^2 - 4 \times 1 \times (-2) = 25 + 8 = 33$.

As the discriminant is positive, the equation has two distinct real roots.

EXAMPLE:

Find the range of values of m such that $mx^2 - 6x + 3 = 0$ has no real roots.

SOLUTION:

The equation has no real roots if $b^2 - 4ac < 0$. Substitute $a = m$, $b = -6$, $c = 3$ into the discriminant.

Hence $b^2 - 4ac < 0 \Rightarrow (-6)^2 - 4 \times m \times 3 < 0 \Rightarrow 36 - 12m < 0 \Rightarrow -12m < -36$
$\Rightarrow m > 3$.

Check all the steps carefully. Remember to reverse the inequality symbol when you divide throughout an inequation by a negative number. If you are unsure, look back to the earlier section on inequations on page 58.

VIDEO LINK

A short recap on 'How to Use the Discriminant' can be found at www.brightredbooks.net/N5Maths

EXAMPLE:

Find the values of p such that the equation $2x^2 - px + 8 = 0$ has equal roots.

SOLUTION:

The equation has equal roots if $b^2 - 4ac = 0$. Substitute $a = 2$, $b = -p$, $c = 8$ into the discriminant.

Hence $b^2 - 4ac = 0 \Rightarrow (-p)^2 - 4 \times 2 \times 8 = 0 \Rightarrow p^2 - 64 = 0$
$$\Rightarrow (p + 8)(p - 8) = 0$$
$$\Rightarrow p = -8 \text{ or } p = 8.$$

BEWARE: We know that when the discriminant is negative, the roots of a quadratic equations are not real. However, if you are asked to solve a quadratic equation correct to 1 decimal place and you then calculate the discriminant to be negative, this would seem unlikely. In this case, check your working carefully.

We have now concluded the section on quadratic functions and equations. Check that you are now aware of the connection between factorisation, parabolas, the quadratic formula and the discriminant.

THINGS TO DO AND THINK ABOUT

1. What can you say about the roots of the equation $x^2 - 2x + 6 = 0$?

2. The equation $x^2 + 6x - p = 0$ has no real roots. Find the range of values of p.

3. Find the values of p such that the equation $x^2 + px + 25 = 0$ has equal roots.

ONLINE TEST

Take the 'The Discriminant' test online at www.brightredbooks.net/N5Maths

PYTHAGORAS' THEOREM AND ITS CONVERSE

PYTHAGORAS' THEOREM

You should be very familiar with Pythagoras' Theorem. There was an example on his Theorem earlier in the section on surds (p. 11). Pythagoras was a Greek mathematician who lived in the 6th century BC. A reminder of his theorem is shown below.

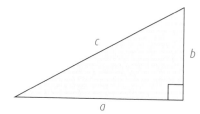

$$a^2 + b^2 = c^2$$

EXAMPLE:

Find x

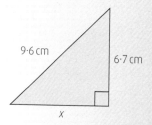

SOLUTION:

Using Pythagoras' Theorem, $9 \cdot 6^2 = 6 \cdot 7^2 + x^2 \Rightarrow x^2 = 9 \cdot 6^2 - 6 \cdot 7^2$

$$= 92 \cdot 16 - 44 \cdot 89$$

$$= 47 \cdot 27.$$

Hence $x = \sqrt{47 \cdot 27} = 6 \cdot 9$. So the length of the side marked x is $6 \cdot 9$ cm (to 2 sig. figs).

EXAMPLES WITH COORDINATES

You can find the distance between two points, given their coordinates, by plotting the points on a grid and then using Pythagoras' Theorem.

EXAMPLE:

Point A has coordinates (–3, 1). Point B has coordinates (4, 3). Find the length of the line AB.

SOLUTION:

Plot A and B and join them. Create a right-angled triangle with AB as the hypotenuse, and then use Pythagoras' Theorem.

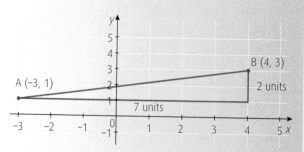

SOLUTION:

$AB^2 = 7^2 + 2^2 = 49 + 4 = 53 \Rightarrow AB = \sqrt{53} = 7 \cdot 3$ units (to 2 sig. figs).

PYTHAGORAS' THEOREM IN THREE DIMENSIONS

EXAMPLE:

AB represents a flagpole at the corner of a field.
The flagpole is 8 metres high. BCDE represents
the field, which is rectangular.
BC is 9 metres long and DC is 24 metres long.

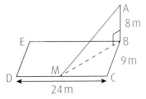

A rope stretches from A, the top of the flagpole, to M, the midpoint of DC.
Calculate the length of the rope.

SOLUTION:

We do two calculations involving Pythagoras' Theorem, firstly in triangle BCM and
secondly in triangle ABM.
$BM^2 = 9^2 + 12^2 = 81 + 144 = 225 \Rightarrow BM = \sqrt{225} = 15$
$AM^2 = 15^2 + 8^2 = 225 + 64 = 289 \Rightarrow AM = \sqrt{289} = 17$, hence the rope is
17 metres long.

VIDEO LINK

Watch the first part of
'Lengths and Angles
inside Cuboids' at www.
brightredbooks.net/N5Maths
for a clear demonstration of
how to find the length of the
space diagonal in a cuboid
using Pythagoras' Theorem.

THE CONVERSE OF PYTHAGORAS' THEOREM

What do we mean by the **converse** of a statement? We find the converse by reversing the
two parts of the statement. So, for a statement 'P→Q', the converse is 'Q→P'. For example,
I might say 'If I live in Glasgow, then I live in Scotland'. This statement is true. The converse
would be 'If I live in Scotland, then I live in Glasgow'. This statement is false. Sometimes the
converse of a statement is true, sometimes false.

Let us consider Pythagoras' Theorem. It can be written as follows:

In a triangle, if angle A is a right angle, then $a^2 = b^2 + c^2$.

The converse is 'In a triangle, if $a^2 = b^2 + c^2$, then angle A is a right angle'.

This converse statement is true and is called the Converse of Pythagoras' Theorem. It is very
useful for proving whether triangles are right-angled or not.

EXAMPLE:

A factory uses a triangular component in one of its machines.

Is the component in the shape of a right-angled triangle?

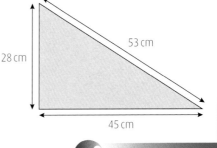

SOLUTION:

Let $a = 53$, $b = 45$, $c = 28$. Now evaluate separately a^2 and $b^2 + c^2$.
Hence $a^2 = 53^2 = 2809$ and $b^2 + c^2 = 45^2 + 28^2 = 2025 + 784 = 2809$.
As $a^2 = b^2 + c^2$, the triangle is right-angled.

**BEWARE: Do not start by substituting into $a^2 = b^2 + c^2$.
Always let a equal the largest side and then evaluate a^2 and
$b^2 + c^2$ separately. If they are equal, you can conclude that
the triangle is right-angled, otherwise it is not.**

ONLINE TEST

Take the 'Pythagoras'
Theorem' test online at
www.brightredbooks.net/
N5Maths

THINGS TO DO AND THINK ABOUT

1. Find the distance between the points C (−1, 2) and D (2, −2).

2. A triangle has sides of length 10 cm, 14 cm and 17 cm. Is it right-angled?

SHAPES AND ANGLES

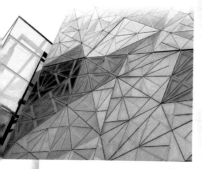

TRIANGLES

Triangles can be scalene (all the sides and angles are different), isosceles (with 2 equal sides and angles) or equilateral (with 3 equal sides and 3 equal angles of 60°). The sum of the angles in a triangle is 180°.

QUADRILATERALS

You should be aware of the following quadrilaterals and their properties:
square; rectangle; rhombus (4 equal sides); parallelogram (opposite sides equal and parallel); kite (2 pairs of equal adjacent sides); trapezium (1 pair of parallel sides).

The sum of the angles in a quadrilateral is 360°.

POLYGONS

The term polygon is usually used to refer to 2-dimensional shapes with more than 4 sides. The table below shows the proper names for some polygons.

Number of sides	5	6	7	8	9	10
Name of polygon	pentagon	hexagon	heptagon	octagon	nonagon	decagon

A regular polygon is a special polygon in which all the sides and angles are equal.

A USEFUL FORMULA

For every regular polygon with n sides, the size of each interior angle, $a°$, is given by the formula

$$a = \frac{180 \times (n-2)}{n}.$$

> **EXAMPLE:**
>
> Find the size of each interior angle of a regular octagon.
>
> **SOLUTION:**
>
> An octagon has 8 sides.
>
> $a = \frac{180 \times (n-2)}{n} = \frac{180 \times (8-2)}{8} = \frac{180 \times 6}{8} = \frac{1080}{8} = 135.$
>
> So, each interior angle of a regular octagon = 135°.

ANGLES

Remember that (i) vertically opposite angles are equal; (ii) corresponding (F-shape) angles are equal for parallel lines; (iii) alternate angles (Z-shape) are equal for parallel lines.

(i)  (ii) 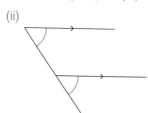 (iii)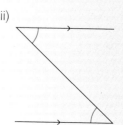

VIDEO LINK

Check you know 'The properties of quadrilaterals' at http://www.youtube.com/watch?v=kJzGlmCGioI

Regular pentagon

VIDEO LINK

A clear video shows how to calculate the sum of the angles in any polygon and leads to the formula given alongside. Watch 'Angles in polygons – interior angles' at www.brightredbooks.net/N5Maths

Regular octagon

THE CIRCLE

The angle in a semi-circle

If AB is a diameter of a circle and C is a point on the circumference of the circle, then angle ACB is a right angle. We normally say that the angle in a semi-circle is a right angle.

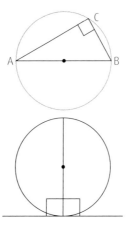

Tangents

A **tangent** to a circle is a straight line which touches the circle at one point only. This point is called the point of contact. At the point of contact, the tangent is perpendicular (at right angles) to the radius (or diameter).

Note also that if two tangents are drawn to a circle from a point outside the circle, then they will be equal in length.

We will now study some angle problems in a circle using the above information. It will help if you remember to watch out for isosceles triangles in circles. Such triangles, with two equal sides and angles, often occur in circles when two radii are drawn.

EXAMPLE:

PQ is a diameter of a circle, centre O. S and R are two points on the circumference of the circle. Angle RPQ = 32° and angle SQP = 25°. Calculate the size of angle RQS.

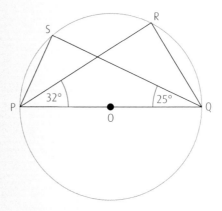

SOLUTION:

$P\hat{R}Q = 90°$ (angle in a semi-circle)

$R\hat{Q}P = 58°$ (sum of angles in a triangle)

$R\hat{Q}S = 33°$ (by subtraction).

Advice

Make sure you know how to name an angle using three letters. Try to set out your working as shown, with a reason for each step. It is recommended that you always copy the diagram and fill in angles on your copy.

ONLINE TEST

Take the 'Shapes and Angles' test online at www.brightredbooks.net/N5Maths

THINGS TO DO AND THINK ABOUT

AB is a tangent to a circle, centre O, with a point of contact at P. Point C lies on the circumference of the circle. Angle APC = 53°.

Calculate the size of angle CDP.

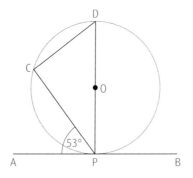

THE CIRCLE

ANGLE PROBLEMS IN A CIRCLE

We started looking at angles in a circle in the previous section. Watch out for angles in a semi-circle and tangents. Use other facts about triangles and straight angles. Always copy diagrams and fill in angles in your diagram, including right angles. Here is another problem.

EXAMPLE:

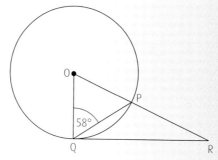

QR is a tangent at point Q to the circle, centre O. Angle OQP = 58°. Calculate the size of angle ORQ.

SOLUTION:

$O\hat{Q}R = 90°$ (angle between radius and tangent)

$O\hat{P}Q = 58°$ (OPQ is an isosceles triangle)

$P\hat{O}Q = 64°$ (sum of angles in triangle OPQ)

$O\hat{R}Q = 26°$ (sum of angles in triangle OQR).

DON'T FORGET

Copy diagrams and fill in as many angles as you need to get the solution.

THE TANGENT-KITE

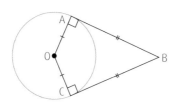

A useful shape to remember is called the tangent-kite. When two tangents are drawn to a circle from a point outside the circle, they are equal in length. Along with the two equal radii, a kite is formed. The angles at A and C are right angles, as they are the angles between tangent and radius.

SYMMETRY IN A CIRCLE

A chord in a circle is a straight line joining two points on the circumference of the circle. A line from the centre of a circle perpendicular to a chord bisects the chord. Similarly, the perpendicular bisector of a chord passes through the centre of a circle.

The presence of right angles in symmetrical diagrams in circles means that many problems can be solved using Pythagoras' Theorem.

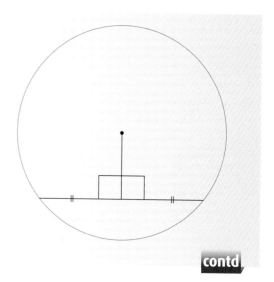

VIDEO LINK

Watch 'Chord Properties/ Perpendicular Bisector' at www.brightredbooks.net/ N5Maths

contd

EXAMPLE:

The entrance to a
tunnel is shaped
like part of a circle.

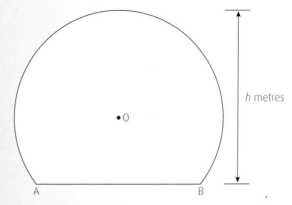

h metres

•O

A   B

The radius of the circle, centre O, is 3·8 metres. The length of the base of the
entrance, represented by AB in the diagram, is 6 metres.

Calculate the height, h metres, of the entrance.

SOLUTION:

This is the type of problem that
can be solved using Pythagoras'
Theorem. Join O to A (or B) and
draw a perpendicular line from O
to the midpoint of AB. This will
form a right-angled triangle. The
perpendicular line will bisect AB.
Let the midpoint of AB be M.
Therefore AM = 3 metres.

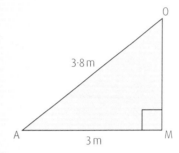

Hence $OM^2 = 3\cdot8^2 - 3^2 = 14\cdot44 - 9 = 5\cdot44 \Rightarrow OM = \sqrt{5\cdot44} = 2\cdot33$.

To find h, the height of the tunnel, we must add the length from O to the top. As
this length is the radius, $h = 3\cdot8 + 2\cdot33 = 6\cdot13$.

Therefore the height of the tunnel is 6·1 metres (to 2 sig. figs).

DON'T FORGET

Many problems concerning
lengths in a circle can be
solved using Pythagoras'
Theorem. It is usually a good
idea to join the centre to the
end of a chord to help make
a right-angled triangle.

ONLINE TEST

Take 'The Circle' test online
at www.brightredbooks.net/
N5Maths

 THINGS TO DO AND THINK ABOUT

A tank is used to transport petrol.

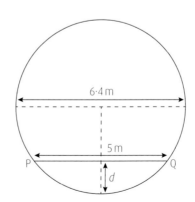

The cross-section of the tank is a circle with diameter 6·4 metres.

The surface of the petrol, represented by PQ in the diagram, is 5 metres wide.

Calculate the depth, d metres, of petrol in the tank.

SIMILAR SHAPES

SIMILAR TRIANGLES

What do we mean in mathematics when we say that two shapes are similar? Think of a slide on an overhead projector and the image of the slide on a screen. They are identical in every respect except that one is an enlargement of the other. In mathematics, similar shapes are like this. We say that two shapes are similar if:

- their corresponding sides are in the same ratio

- they are equiangular.

In the special case of triangles, if one of these conditions is true, then it follows that the other will also be true. Consider the triangles shown below in which AB is parallel to DE.

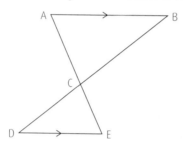

Angles CAB and CED are equal. Angles ABC and CDE are equal. Both pairs are alternate angles (Z-shapes). The angles at C (ACB and DCE) are equal vertically opposite angles. Hence the triangles are equiangular, that is they have equal angles. Hence they are similar and their corresponding sides must be in the same ratio, that is

$$\frac{AB}{DE} = \frac{AC}{CE} = \frac{BC}{CD}.$$

The value of each ratio is called the **scale factor**. A scale factor > 1 leads to an enlargement, and a scale factor between 0 and 1 leads to a reduction.

VIDEO LINK

Watch 'Similar Triangles' at www.brightredbooks.net/N5Maths for a clear explanation.

DON'T FORGET

It may help if you draw the two triangles separated (as shown in the diagram).

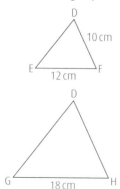

EXAMPLE:

In the diagram, EF is parallel to GH.

EF = 12 centimetres, DF = 10 centimetres and GH = 18 centimetres.

Calculate the length of FH.

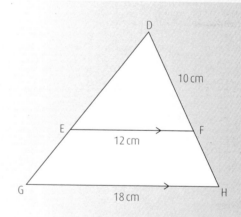

SOLUTION:

Consider triangles DEF and DGH.

$\hat{E} = \hat{G}$ and $\hat{F} = \hat{H}$ (corresponding angles).

As $\hat{D}$ is a common angle, triangles are equiangular and hence similar.

We can find the scale factor from the ratio $\frac{GH}{EF}$. As $\frac{GH}{EF} = \frac{18}{12}$, the scale factor is 1·5, that is, the lengths in triangle DGH are 1·5 times larger than those in triangle DEF. As side DH corresponds to side DF, then DH is 1·5 × 10 = 15 centimetres.

Hence FH = 15 – 10 = 5 centimetres.

SIMILAR SHAPES AND THEIR AREAS

In the same way as lengths are measured in centimetres and area is measured in square centimetres, we must use the idea of squaring to find areas involved with similar shapes. If two shapes are similar, with a length scale factor of f, then the area scale factor is f^2.

EXAMPLE:

A furniture company has produced a set of octagonal tables.

The tops of the tables are mathematically similar.

 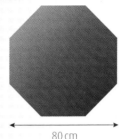

64 cm

80 cm

The area of the top of the smaller table is 3393 square centimetres.

Calculate the area of the top of the larger table.

SOLUTION:

The length scale factor is $\frac{80}{64}$ = 1·25

The area scale factor is $1·25^2$

The area of the top of the larger table is $1·25^2 \times 3393$ = 5302 cm²

SIMILAR SHAPES AND THEIR VOLUMES

In the same way as lengths are measured in centimetres and volume is measured in cubic centimetres, we must use the idea of cubing to find volumes involved with similar shapes. If two shapes are similar, with a length scale factor of f, then the volume scale factor is f^3.

EXAMPLE:

Customers in a café can buy two different-sized cups of coffee, 'small' and 'large'. The two cups are mathematically similar. The small cup can hold 400 millilitres and is 10 centimetres high. The large cup is 12 centimetres high.

Calculate the volume of the large cup.

SOLUTION:

The length scale factor is $\frac{12}{10}$ = 1·2

The volume scale factor is $1·2^3$

The volume of the large cup is $1·2^3 \times 400$ = 691·2 ml.

 DON'T FORGET

Remember that if the length scale factor is f, then the area scale factor is f^2, and the volume scale factor is f^3.

VIDEO LINK

There is more explanation of this in 'Scale Factor, Length, Area and Volume' at www.brightredbooks.net/N5Maths

 THINGS TO DO AND THINK ABOUT

1. Two television screens are mathematically similar. The first screen has a diagonal length of 25 inches and an area of 275 square inches. The second screen has a diagonal length of 30 inches. Find the area of the second screen.

2. Two milk cartons are mathematically similar. The smaller carton holds 1 litre and is 20 centimetres high. The larger carton is 22 centimetres high. What is the volume of the larger carton?

 ONLINE TEST

Take the 'Similar Shapes' test online at www.brightredbooks.net/N5Maths

TRIGONOMETRIC GRAPHS

BASIC GRAPHS

You should know the basic graphs of $y = \sin x°$, $y = \cos x°$ and $y = \tan x°$.

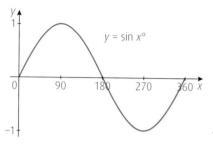 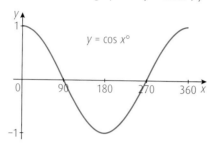 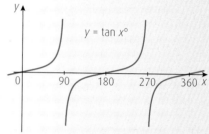

The above graphs are periodic, that is they have a repeating pattern. We say that the period of a graph is the length of one cycle. The periods of the three graphs shown are 360° for $y = \sin x°$ and $y = \cos x°$ and 180° for $y = \tan x°$.

AMPLITUDE

The graphs of $y = k \sin x°$ and $y = k \cos x°$, where k is a constant, change the maximum and minimum values of the basic sine and cosine graphs from 1 and −1 to k and $−k$. In other words, the graphs are stretched vertically. The value k is called the amplitude of the graph. Note that the period of these graphs is unchanged.

> **EXAMPLE:**
>
> Draw the graph of $y = 3 \sin x°$, $0 \leqslant x \leqslant 360$.
>
> **SOLUTION:**

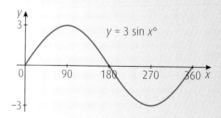

VERTICAL TRANSLATION

The graphs of $y = \sin x° + k$ and $y = \cos x° + k$, where k is a constant, slide the sine and cosine graphs up or down k units depending on whether k is positive or negative. The period of these graphs is again unchanged. This way of moving a graph is called a vertical translation. In this case, every point on a graph moves in the same direction (vertically) and by the same distance.

> **EXAMPLE:**
>
> Draw the graph of $y = \sin x° + 2$, $0 \leqslant x \leqslant 360$.
>
> **SOLUTION:**

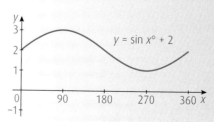

MULTIPLE ANGLES

In the graphs of $y = \sin kx°$ and $y = \cos kx°$, where k is a constant, the value of k tells you how many cycles of the basic sine or cosine graphs appear between 0° and 360°. In both of these cases, the period is altered to $(360 ÷ k)°$.

EXAMPLE:

Draw the graph of $y = \sin 2x°$, $0 \leqslant x \leqslant 360$.

SOLUTION:

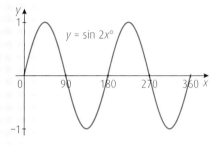

NOTE: The period of this graph is $(360 ÷ 2)° = 180°$.

EXAMPLE:

What is the period of $y = 5 \sin 3x°$, $0 \leqslant x \leqslant 360$?

SOLUTION:

The basic period of $y = \sin x°$ is 360°. The period of $y = 5 \sin 3x°$ is unchanged by the amplitude (5) but is altered by the multiple angle to $(360 ÷ 3)° = 120°$.

VIDEO LINK

For an entertaining musical and visual take on 'Trigonometric Graphs', watch www.brightredbooks.net/N5Maths

EXAMPLE:

The graph of $y = a \sin bx°$, $0 \leqslant x \leqslant 360$, is shown in the graph alongside.
State the values of a and b.

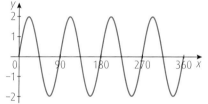

SOLUTION:

The value of a is 2 (the amplitude). The value of b is 4 (the number of cycles of the basic sine graph appearing between 0° and 360°).

DON'T FORGET

Check carefully all these ideas (the basic graphs, amplitude, vertical translation, multiple angle and phase angle) until you understand them.

PHASE ANGLE

We next look at trigonometric functions such as $y = \sin (x - a)°$ and $y = \cos (x - a)°$. These functions involve a phase angle, $a°$, which has the effect of moving the basic sine and cosine graphs $a°$ to the right. The period of the basic graphs is unchanged.

EXAMPLE:

Draw the graph of $y = \sin (x - 45)°$, $0 \leqslant x \leqslant 360$.

SOLUTION:

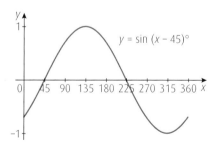

NOTE: Did you notice that if x is replaced by 45 in the equation $y = \sin (x - 45)°$, you get $y = \sin 0°$ which is equal to 0? This helps explain why the graph of $y = \sin (x - 45)°$ is the same as the graph of $y = \sin x°$ moved 45° to the right.

Trigonometric functions such as $y = \sin (x + a)°$ and $y = \cos (x + a)°$ have the effect of moving the basic sine and cosine graphs $a°$ **to the left**.

ONLINE TEST

Take the 'Trigonometric Graphs' test online at www.brightredbooks.net/N5Maths

 THINGS TO DO AND THINK ABOUT

Sketch the graph of the function $y = 4 \cos 2x°$, $0 \leqslant x \leqslant 360$, showing the maximum and minimum values of the function and the points where the graph crosses the x-axis.

RELATED ANGLES

THE RIGHT-ANGLED TRIANGLE

You should remember the basic trigonometric ratios in a right-angled triangle.

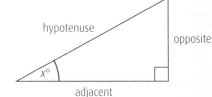

$$\tan x° = \frac{\text{opposite}}{\text{adjacent}}$$

$$\sin x° = \frac{\text{opposite}}{\text{hypotenuse}}$$

$$\cos x° = \frac{\text{adjacent}}{\text{hypotenuse}}$$

EXAMPLE:

Find the length of the side marked x in the diagram.

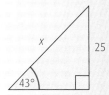

SOLUTION:

$$\sin 43° = \frac{25}{x} \Rightarrow x \sin 43° = 25 \Rightarrow x = \frac{25}{\sin 43°} = 36{\cdot}7$$

In the above example, we used sin 43°. If you check your calculator, you will see that sin 43° = 0·681 998 360 1. If you key sin 137° into your calculator, you will see the answer is identical. We shall look at how these angles are related and other connections between the sine, cosine and tangent of angles between 0° and 360°.

RELATED ANGLES

We shall consider the basic sine, cosine and tangent graphs from the previous section.

We can see from the symmetry of the sine graph that sin 43° = sin 137° = 0·682 (correct to 3 sig. figs). If the sine graph was extended in both directions, a whole series of angles arise whose sines are equal.

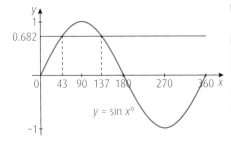

Did you notice that 43° and 137° are supplementary angles, that is they add up to 180°? By considering the graph, you can also notice that the sine of every angle between 0° and 180° is positive and the sine of every angle between 180° and 360° is negative. If we study the basic trigonometric graphs, we also notice that something important happens at the multiples of 90°. Either the graphs have a maximum or minimum turning point, or they cross the x-axis. By thinking of the interval between 0° and 360° being divided into four quarters (or **quadrants**), we can see how the values of the three trigonometric ratios vary between positive and negative. Check the results below against the three graphs.

Quadrant	1st	2nd	3rd	4th
Angle size	0° → 90°	90° → 180°	180° → 270°	270° → 360°
Sine	Positive (+)	Positive (+)	Negative (−)	Negative (−)
Cosine	Positive (+)	Negative (−)	Negative (−)	Positive (+)
Tangent	Positive (+)	Negative (−)	Positive (+)	Negative (−)

This information is more usually described in a grid in which we think of positive angles as being measured anti-clockwise from the x-axis:

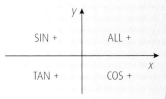

contd

Note that the following important values can be read off from the basic graphs:

sin 0° = 0, sin 90° = 1, sin 180° = 0, sin 270° = –1, sin 360° = 0

cos 0° = 1, cos 90° = 0, cos 180° = –1, cos 270° = 0, cos 360° = 1.

When we consider angles greater than 90°, we can find a related acute angle. We saw that 137° was related to 43° (180° – 43°). By considering the basic graphs, we can find the connections which relate all angles to an acute angle. This becomes essential, along with the information in the CAST diagram, when we go on to solve trigonometric equations.

For an angle A in the 1st quadrant:

- the related angle in the 2nd quadrant is (180° – A)

- the related angle in the 3rd quadrant is (180° + A)

- the related angle in the 4th quadrant is (360° – A).

The following examples would be tested without a calculator, so do not use a calculator.

DON'T FORGET

You must remember the information about when the trigonometric ratios are positive or negative. Use either 'all, sin, tan, cos' or a mnemonic such as CAST.

EXAMPLE:
Given that sin 30° = 0·5, what is the value of sin 330°?

SOLUTION:
First, check the CAST diagram to see whether the solution is positive or negative. This tells us that sin 330° is negative, as 330° is in the 4th quadrant. Then express 330° in terms of its related acute angle.

Hence sin 330° = sin (360 – 30)° = –sin 30° = –0·5.

A common error
Some students divide 330 by 30 (= 11) and assume that sin 330° must be 11 × sin 30° = 11 × 0·5 = 5·5. This is completely wrong, so avoid at all costs.

DON'T FORGET

Always use 180° or 360° when finding a related angle. Never use 90° or 270° when finding a related angle.

EXAMPLE:
Given that tan 60° = √3, what is the value of tan 240°?

SOLUTION:
As 240° is in the third quadrant, tan 240° is positive. Now use the related angle.
Hence tan 240° = tan (180 + 60)° = +tan 60° = √3.

EXAMPLE:
Which of the following has the greatest value?
sin 190°, cos 152°, tan 265°.
You must give a reason for your answer.

SOLUTION:
By considering the CAST diagram, you can tell that sin 190° and cos 152° are negative and tan 265° is positive, therefore tan 265° has the greatest value.

VIDEO LINK

Some students may have been taught related angles and the CAST diagram with reference to angles on a coordinate grid. You can watch a demonstration of this online by watching 'The CAST Rule. Signs of the Ratios: Trigonometry Math Help' at www.brightredbooks. net/N5Maths

ONLINE TEST

Take the 'Related Angles' test online at www. brightredbooks.net/N5Maths

THINGS TO DO AND THINK ABOUT

1. Given that sin 30° = 0·5, what is the value of sin 150°?

2. Given that tan 35° = 0·7, find another angle whose tangent is 0·7 in the interval 0 ⩽ x ⩽ 360.

TRIGONOMETRIC EQUATIONS

SOLVING BASIC EQUATIONS

DON'T FORGET

Make sure your calculator is in the correct mode, that is, DEGREES and not RAD or GRAD. Do a simple check: for example, tan 45° = 1 is easy to remember.

It is expected that you will be able to solve basic trigonometric equations in the interval $0 \leqslant x \leqslant 360$. It is important that you remember that most of the equations you will be asked to solve will have two solutions. In order to do this, you will need to refer to the information from the previous section on related angles and the CAST diagram. This is summarised below.

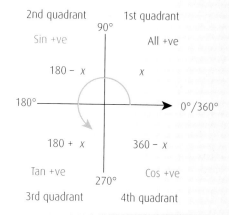

EXAMPLE:

Solve the equation $\cos x° = 0.548$, $0 \leqslant x \leqslant 360$.

SOLUTION:

Decide in which two quadrants cosine is positive, that is, 1st and 4th.

Use inverse cosine on your calculator to find $x = 56.8$, then use related angles for the 4th quadrant, giving $360 - 56.8 = 303.2$. Hence $x = 56.8$ or 303.2.

NOTE: You can check your solutions by evaluating $\cos 56.8°$ and $\cos 303.2°$ on your calculator. The values should round to 0.548.

When the value of a trigonometric ratio is negative, solving equations is more difficult. Study the following example very carefully.

DON'T FORGET

When solving an equation with a negative trigonometric ratio, you need to find the related acute angle in the 1st quadrant. This can be done by omitting the negative sign. Remember also to check your solutions.

EXAMPLE:

Solve the equation $\tan x° = -2.4$, $0 \leqslant x \leqslant 360$.

SOLUTION:

Decide in which two quadrants tangent is negative, that is, 2nd and 4th.

Although it is not a solution, we need to find the related acute angle in the 1st quadrant. Therefore, use inverse tangent on your calculator to find $\tan^{-1} 2.4 = 67.4$ (you should omit the negative sign). Now use this value to find the two solutions in the 2nd and 4th quadrants.

Hence $x = 180 - 67.4 = 112.6$ and $360 - 67.4 = 292.6$.

MORE DIFFICULT EQUATIONS

In some equations, we will have to do some rearranging before we proceed to the final solution. Equations should be in the form $\sin x° = \ldots$, $\cos x° = \ldots$ or $\tan x° = \ldots$ before arriving at the final solution.

Consider the following example.

contd

EXAMPLE:

Solve the equation 5 sin $x°$ + 3 = 1, 0 ⩽ x ⩽ 360.

SOLUTION:

5 sin $x°$ + 3 = 1

$\Rightarrow$ 5 sin $x°$ = 1 − 3   (subtract 3 from both sides)

$\Rightarrow$ 5 sin $x°$ = −2

$\Rightarrow$ sin $x°$ = $-\frac{2}{5}$ = −0·4   (divide both sides by 5).

As sine is negative, solutions are in the 3rd and 4th quadrants.

Now find the related angle in the 1st quadrant: sin^{-1} 0·4 = 23·6.

Hence x = 180 + 23·6 = 203·6 or x = 360 − 23·6 = 336·4.

SUMMARY

When solving basic trigonometric equations:

- arrange in the form sin $x°$ = ..., cos $x°$ = ... or tan $x°$ = ...
- use the CAST diagram to decide which two quadrants apply
- if there is a negative ratio, omit the negative sign when finding the related acute angle
- use 180° or 360° to find solutions in the required quadrants.

VIDEO LINK

To see how trigonometric equations can be solved by reference to the basic trigonometric graphs, watch 'No. 21 MORE Trig' at www.brightredbooks.net/N5Maths

PROBLEMS INVOLVING TRIGONOMETRIC EQUATIONS

EXAMPLE:

As the arms on a wind turbine rotate, the height, h metres, of the tip of one of the arms is given by the equation:

$$h = 9 + 5 \sin t°$$

where t is the time in seconds after the turbine is switched on.

(a) Calculate the height of the tip of the arm after 20 seconds.

(b) Find **two** times during the first turn of the arms when the tip of the arm is 12 metres high.

SOLUTION:

(a) h = 9 + 5 sin $t°$ = 9 + 5 sin 20° = 10·7 metres.

(b) h = 9 + 5 sin $t°$ $\Rightarrow$ 12 = 9 + 5 sin $t°$

$\Rightarrow$ 5 sin $t°$ = 12 − 9

$\Rightarrow$ 5 sin $t°$ = 3

$\Rightarrow$ sin $t°$ = $\frac{3}{5}$ = 0·6.

Solutions are in the 1st and 2nd quadrants.

Hence t = 36·9 or x = 180 − 36·9 = 143·1

Therefore the arm is 12 metres high after 36·9 and 143·1 seconds.

 THINGS TO DO AND THINK ABOUT

Solve the following equations

(a) 4 sin $x°$ − 3 = − 2, 0 ⩽ x ⩽ 360   (b) 2 cos $x°$ + 1 = 2, 0 ⩽ x ⩽ 360

(c) 5 sin $x°$ + 4 = 1, 0 ⩽ x ⩽ 360   (d) 8 tan $x°$ + 1 = 6, 0 ⩽ x ⩽ 360

(e) 3 tan $x°$ + 5 = − 2, 0 ⩽ x ⩽ 360   (f) 4 cos $x°$ − 2 = − 5, 0 ⩽ x ⩽ 360.

ONLINE TEST

Take the 'Trigonometric Equations' test online at www.brightredbooks.net/N5Maths

TRIGONOMETRIC IDENTITIES

WHAT IS A TRIGONOMETRIC IDENTITY?

VIDEO LINK

For a clear proof of the first of these formulae, watch 'Trigonometric Identities Are Beautiful' at www.brightredbooks.net/N5Maths

Basically, a trigonometric identity is an equation that is true for all values of the variables. You will be asked to prove or show that an identity is true. To do this, you will have to use two identities which involve sine, cosine and tangent.

$$\sin^2 A + \cos^2 A = 1 \text{ and } \tan A = \frac{\sin A}{\cos A}$$

Note that $\sin^2 A$ and $\cos^2 A$ are the same as $(\sin A)^2$ and $(\cos A)^2$. These two identities are true no matter what value A takes. For example, suppose we choose a value of 65° for A. If you evaluate $\sin^2 65° + \cos^2 65°$ on your calculator, you will get a solution of 1. In fact, no matter what value you choose for A, the solution is always 1. The same thing applies to the second formula involving tangent. This explains why such equations are called identities.

It is important that you realise that substituting a particular value of your choice into one side of an identity and verifying that it equals the other side does not count as being a proof. To prove that an equation is a trigonometric identity requires much more than simply substituting in a few values.

USING TRIGONOMETRIC IDENTITIES

You should be familiar with the above formulae; and we will look at types of examples where they can prove useful. As this topic is usually tested without a calculator, do not use a calculator from now on.

EXAMPLE:

If $\sin A = \frac{3}{5}$, $0 \leqslant A \leqslant 90$, find the values of (a) $\cos A$; (b) $\tan A$.

SOLUTION:

(a) $\sin^2 A + \cos^2 A = 1 \Rightarrow \left(\frac{3}{5}\right)^2 + \cos^2 A = 1 \Rightarrow \cos^2 A = 1 - \left(\frac{3}{5}\right)^2 = 1 - \frac{9}{25} = \frac{16}{25}$
Hence $\cos A = \sqrt{\frac{16}{25}} = \frac{4}{5}$.

(b) $\tan A = \frac{\sin A}{\cos A} = \frac{3}{5} \div \frac{4}{5} = \frac{3}{5} \times \frac{5}{4} = \frac{3}{5}^1 \times \frac{5^1}{4} = \frac{3}{4}$.

Because $0 \leqslant A \leqslant 90$, we do not have to include the negative square root in part (a), as cosine is positive in this interval. Note that there are other methods of solving the above problem using trigonometry in a right-angled triangle (SOHCAHTOA) and Pythagoras' Theorem.

DON'T FORGET

Memorise the two trigonometric identities given at the beginning of this section and the rearrangements given here.

In part (a), we rearranged the identity $\sin^2 A + \cos^2 A = 1$ during the process of arriving at a solution. By rearranging this formula, we can get two useful variations of the formula which we will need for other proofs to follow.

$\sin^2 A + \cos^2 A = 1$ leads to

$\sin^2 A = 1 - \cos^2 A$ and

$\cos^2 A = 1 - \sin^2 A$.

PROOFS

You will often be asked in mathematics to prove something or to show that two expressions are equal.

It is common to be asked to prove that a trigonometric identity is true. Such proofs are done in a special way, by starting with one side of the equation and manipulating it until you show that it is equal to the other side. You can start with either the right or the left side, although mostly you will start with the left side.

contd

Abbreviations

Use LS for left side (or LHS for left-hand side) and use RS for right side (or RHS for right-hand side). You then aim to prove that LS = RS.

> **EXAMPLE:**
>
> Prove that $\tan^2 x° \cos^2 x° = \sin^2 x°$.

> **SOLUTION:**
>
> $LS = \tan^2 x° \cos^2 x°$
>
> $= \left(\dfrac{\sin x°}{\cos x°}\right)^2 \times \cos^2 x°$
>
> $= \dfrac{\sin^2 x°}{\cos^2 x°} \times \cos^2 x°$
>
> $= \sin^2 x°$
>
> $= RS$.
>
> NOTE: All trigonometric identities should be solved using this type of format: that is, start with one side (LS) and eventually prove that it equals the other side (= RS). To carry out the proof, you will need to use the trigonometric identities given earlier in the section. In the above example, remember that $\tan^2 x° = (\tan x°)^2$.

DON'T FORGET

Never start to prove an identity by copying the full equation. It is up to you to prove that the two sides are equal, so always start with one side only.

> **EXAMPLE:**
>
> Prove that $\sin x° \cos^2 x° + \sin^3 x° = \sin x°$.

> **SOLUTION:**
>
> When you inspect the LS, the only part associated with one of the earlier identities is $\cos^2 x°$. If you remember that $\cos^2 A = 1 - \sin^2 A$, you can make a substitution.
>
> Hence $LS = \sin x° \cos^2 x° + \sin^3 x°$
>
> $= \sin x° (1 - \sin^2 x°) + \sin^3 x°$
>
> $= \sin x° - \sin^3 x° + \sin^3 x°$
>
> $= \sin x°$
>
> $= RS$.

The best way to become efficient at proving trigonometric identities is to study the format for doing so, learn the important formulae and then practise a number of examples. Finally, you may be asked to simplify a trigonometric expression using the formulae.

> **EXAMPLE:**
>
> Simplify $\dfrac{\sin^3 x°}{1 - \cos^2 x°}$.

> **SOLUTION:**
>
> $\dfrac{\sin^3 x°}{1 - \cos^2 x°} = \dfrac{\sin^3 x°}{\sin^2 x°} = \sin x°$.
>
> NOTE: The denominator also could have been simplified by replacing 1 with $\sin^2 x° + \cos^2 x°$ from the first formula in the section.

ONLINE TEST

Take the 'Trigonometric Identities' test online at www.brightredbooks.net/N5Maths

THINGS TO DO AND THINK ABOUT

1. Prove that $4\cos^2 x° + 5\sin^2 x° = 4 + \sin^2 x°$.

2. Show that $\sin x° \cos x° \tan x° = \sin^2 x°$.

APPLICATIONS

THE AREA OF A TRIANGLE

REVISION

You are already familiar with the formula $A = \frac{1}{2}bh$ for the area of a triangle, where b is the base and h is the height.

> **EXAMPLE:**
>
> Find the area of triangle ABC.
>
> **SOLUTION:**
>
> $A = \frac{1}{2}bh$
>
> $\quad = \frac{1}{2} \times 8 \times 6 \cdot 2$
>
> $\quad = 24 \cdot 8$
>
> Hence area of triangle = 24·8 cm².
>
> Note that the height can be inside the triangle (as shown above), a side of the triangle (in a right-angled triangle) or outside the triangle (in an obtuse-angled triangle).

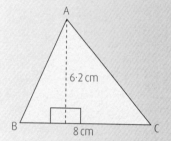

A NEW FORMULA

We will now learn another formula for finding the area of a triangle, involving trigonometry. Before doing that, you must know the rules for naming angles and sides in a triangle. We use capital letters to name angles, for example A, B, C, and italic small letters to name sides, for example a, b, c, where a is the side opposite angle A, that is side BC in triangle ABC.

Consider the area of triangle ABC:

$\sin C = \frac{h}{b} \Rightarrow h = b \sin C$

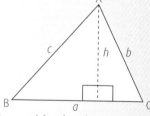

Area of triangle ABC is $\quad A = \frac{1}{2} \times$ base $\times$ height

$$A = \frac{1}{2}ab \sin C.$$

It is important that you realise that different variables can be used for the above area formula depending on the given information. Three versions of the same formula are given below.

$A = \frac{1}{2}ab \sin C \quad$ or $\quad A = \frac{1}{2}bc \sin A \quad$ or $\quad A = \frac{1}{2}ac \sin B.$

This formula can therefore be adapted to suit different variables as required. To use this formula, you will need to be given the lengths of two sides in a triangle and the size of the included angle, that is the angle in between the two given sides.

 DON'T FORGET

Use this triangle formula when you know two sides and the included angle.

> **EXAMPLE:**
>
> The sketch shows a triangle, PQR.
>
> Calculate the area of the triangle.
>
> **SOLUTION:**
>
> $A = \frac{1}{2}pr \sin Q = \frac{1}{2} \times 13 \times 15 \times \sin 83° = 96 \cdot 773$
>
> Hence area of triangle is 97 cm² (correct to 2 sig. figs).

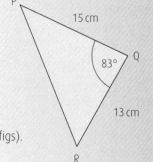

WORKING BACK

We can work back from knowing the area to find the length of a side or the size of an angle. In the following example, we shall do this by changing the subject of the formula.

> **EXAMPLE:**
>
> In triangle DEF, angle EDF = 72°.
>
> DE = 24 centimetres.
>
> The area of triangle DEF is 205 square centimetres.
>
> Calculate the length of side DF.

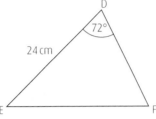

> **SOLUTION:**
>
> $A = \frac{1}{2}ef \sin D \Rightarrow 2A = ef \sin D \Rightarrow e = \frac{2A}{f \sin D}$
>
> $\Rightarrow e = \frac{2 \times 205}{24 \times \sin 72°} = \frac{410}{22\cdot825} = 17\cdot96$
>
> Hence the length of DF is 18 cm (correct to 2 sig. figs).

AREA OF QUADRILATERALS

By drawing a diagonal in a quadrilateral, it is split into two triangles. The area of the quadrilateral can then be found by adding the area of the triangles together.

> **EXAMPLE:**
>
> WXYZ is a rhombus of side 8 centimetres.
>
> Angle XYZ = 140°.
>
> Calculate the area of the rhombus.

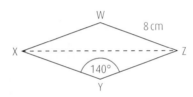

> **SOLUTION:**
>
> Area of triangle WXZ $= \frac{1}{2}xz \sin W$
>
> $= \frac{1}{2} \times 8 \times 8 \times \sin 140°$
>
> $= 20\cdot57$
>
> Area of rhombus = 2 × 20·57 = 41·14 (due to symmetry of rhombus)
>
> Hence area = 41 cm² (correct to 2 sig. figs).

VIDEO LINK

You can move triangles about and check their areas using the above formula at www. brightredbooks.net/N5Maths

A NON-CALCULATOR EXAMPLE

> **EXAMPLE:**
>
> The area of triangle ABC is 80 square centimetres.
>
> AB = 20 centimetres and $\sin B = \frac{4}{5}$.
>
> Calculate the length of BC.

> **SOLUTION:**
>
> $A = \frac{1}{2}ac \sin B \Rightarrow 80 = \frac{1}{2} \times a \times 20 \times \frac{4}{5} \Rightarrow 80 = 8a \Rightarrow a = 10$
>
> Hence BC = 10 cm.

ONLINE TEST

Take the 'Area of a Triangle' test online at www. brightredbooks.net/N5Maths

 THINGS TO DO AND THINK ABOUT

Find the area of the triangle shown.

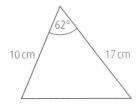

THE SINE RULE

WHEN TO USE THE SINE RULE

In mathematics, we can use the ideas from trigonometry (SOHCAHTOA) and Pythagoras' Theorem to solve many problems in right-angled triangles. However, in triangles which are not right-angled, we need other formulae to find the lengths of sides and the sizes of angles. One of the most important formulae is called the **sine rule**. The formula for triangle ABC is given below.

$$\frac{a}{\sin A} = \frac{b}{\sin B} = \frac{c}{\sin C}$$

This formula can be used in *any* triangle

- to find the length of a side when you are given two angles and one other side
- to find the size of an angle when you are given two sides and an angle *other than* the included angle.

FINDING THE LENGTH OF A SIDE

EXAMPLE:

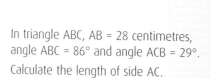

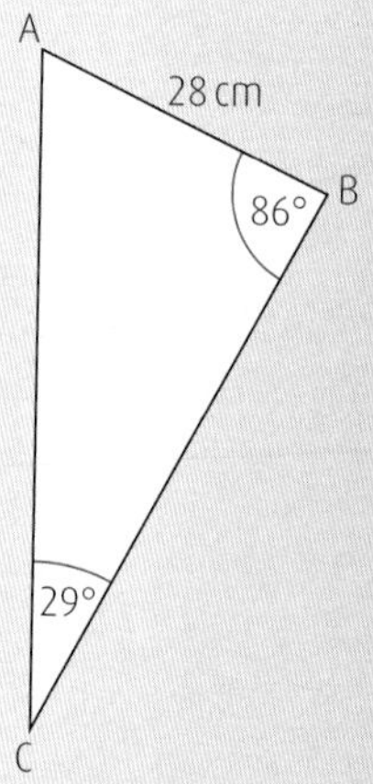

In triangle ABC, AB = 28 centimetres, angle ABC = 86° and angle ACB = 29°.

Calculate the length of side AC.

SOLUTION:

$$\frac{a}{\sin A} = \frac{b}{\sin B} = \frac{c}{\sin C} \Rightarrow \frac{b}{\sin 86°} = \frac{28}{\sin 29°} \quad \Rightarrow b \times \sin 29° = 28 \times \sin 86°$$

$$\Rightarrow b = \frac{28 \times \sin 86°}{\sin 29°} = 57{\cdot}61$$

Hence AC = 58 cm (correct to 2 sig. figs).

DON'T FORGET

Once you have substituted into the sine rule, choose the two relevant equal ratios, then cross-multiply to continue the calculation.

EXAMPLE:

For triangle ABC, shown above, calculate the length of side BC.

SOLUTION:

At first, this appears impossible, as we have no information about A (the angle) or *a* (the side we have been asked to find). However, it is easy to calculate the size of angle A, as the sum of the angles in a triangle is 180°.

Hence angle BAC = (180 − 86 − 29)° = 65°.

Now we can use the sine rule:

$$\frac{a}{\sin A} = \frac{b}{\sin B} = \frac{c}{\sin C} \Rightarrow \frac{a}{\sin 65°} = \frac{28}{\sin 29°} \Rightarrow a \times \sin 29° = 28 \times \sin 65°$$

$$\Rightarrow a = \frac{28 \times \sin 65°}{\sin 29°} = 52{\cdot}34$$

Hence BC = 52 cm (correct to 2 sig. figs).

DON'T FORGET

In trigonometry problems, when you know the size of two angles in a triangle, you can always find the third angle, as all three angles add up to 180°.

FINDING THE SIZE OF AN ANGLE

EXAMPLE:

In triangle ABC, AC = 22 centimetres,
BC = 28 centimetres, and angle BAC = 59°.

Calculate the size of angle ABC.

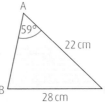

SOLUTION:

$\frac{a}{\sin A} = \frac{b}{\sin B} = \frac{c}{\sin C} \Rightarrow \frac{28}{\sin 59°} = \frac{22}{\sin B} \Rightarrow 28 \times \sin B = 22 \times \sin 59°$

$\Rightarrow \sin B = \frac{22 \times \sin 59°}{28} = 0.673$

Hence angle ABC = $\sin^{-1} 0.673 = 42.3°$.

Although this problem has been solved, there is potentially a difficulty. If you think back to the section on trigonometric equations, you will remember that there are *two* possible solutions to the equation sin B = 0.673, both of which could occur in a triangle. By thinking of the CAST diagram and using related angles, we find that x = 42.3 or 180 – 42.3, that is 42.3 or 137.7. This arises because sine is positive in the first and second quadrants. However, we can see that angle B is an acute angle, hence angle ABC = 42.3°.

You should be aware that occasionally the solution will be an obtuse angle. This could arise if the triangle you are dealing with is obtuse-angled. If not, the solution will be in the first quadrant and will be accessed directly from your calculator. If the triangle is obtuse-angled, remember that the obtuse angle will always be opposite the *longest* side in the triangle.

 DON'T FORGET

In any triangle, the longest side is opposite the largest angle, and the shortest side is opposite the smallest angle.

A NON-CALCULATOR EXAMPLE

EXAMPLE:

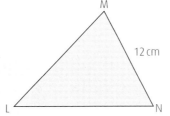

In triangle LMN, MN = 12 centimetres, sin M = $\frac{2}{3}$ and sin L = $\frac{1}{2}$.

Calculate the length of side LN.

SOLUTION:

$\frac{l}{\sin L} = \frac{m}{\sin M} = \frac{n}{\sin N} \Rightarrow \frac{12}{\frac{1}{2}} = \frac{m}{\frac{2}{3}} \Rightarrow 12 \times \frac{2}{3} = m \times \frac{1}{2} \Rightarrow 8 = m \times \frac{1}{2} \Rightarrow m = 8 \div \frac{1}{2}$

$\Rightarrow m = 8 \times \frac{2}{1} = 16$

Hence LN = 16 cm.

NOTE: Examples without a calculator can be tricky because fractions are involved. We will shortly have a more in-depth look at fractions, however. Avoid a common mistake of writing sin $\frac{2}{3}$ rather than simply $\frac{2}{3}$ when substituting in the sine rule for sin M.

 VIDEO LINK

Watch a clear clip on finding the size of an angle: 'Finding Angles Using the Sine Rule' at www.brightredbooks.net/N5Maths

 ONLINE TEST

Take the test 'The Sine Rule' online at www.brightredbooks.net/N5Maths

 ## THINGS TO DO AND THINK ABOUT

In the two examples below, make a rough sketch of the triangles to help get started on the formula.

1. In triangle ABC, angle A = 48°, angle B = 63° and AC = 5 metres. Calculate BC.

2. In triangle RST, angle S = 59°, RT = 10 metres and RS = 8 metres. Calculate angle RTS.

THE COSINE RULE

WHEN TO USE THE COSINE RULE

In the previous section, we found out when to use the sine rule. Depending on the information we are given in a triangle, it is not always possible to use the sine rule. For example, if we are told the lengths of two sides and the included angle in a triangle and are asked to find the length of the third side, it is clear that the sine rule cannot be used. In this case, we use the **cosine rule**. The formula for triangle ABC is given below.

$$a^2 = b^2 + c^2 - 2bc \cos A$$

This formula can be used in *any* triangle to find the length of a side when you are given two sides and the included angle.

It is important that you realise that different variables can be used for the above formula depending on the given information. Three versions of the same formula are given below.

$$a^2 = b^2 + c^2 - 2bc \cos A \quad \text{or} \quad b^2 = a^2 + c^2 - 2ac \cos B \quad \text{or} \quad c^2 = a^2 + b^2 - 2ab \cos C$$

DON'T FORGET

The included angle is the angle between the two given sides. This is the same information that is used to find the area of a triangle.

USING THE COSINE RULE TO FIND THE LENGTH OF A SIDE

EXAMPLE:

In triangle ABC, angle BAC = 78°.
AB = 36 centimetres.
AC = 23 centimetres.
Calculate the length of side BC.

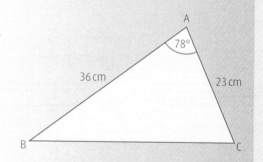

SOLUTION:

$$a^2 = b^2 + c^2 - 2bc \cos A$$
$$= 23^2 + 36^2 - 2 \times 23 \times 36 \times \cos 78°$$
$$= 1480{\cdot}698\,24$$

Hence $a = \sqrt{1480{\cdot}698\,24} = 38{\cdot}48$, so BC = 38 cm (correct to 2 sig. figs).

Advice

Examples like this will be solved using a calculator, so there is no need to split up the calculation of $23^2 + 36^2 - 2 \times 23 \times 36 \times \cos 78°$ into individual parts. The whole calculation can all be carried out in one go on your calculator. You should practise this to gain confidence.

DON'T FORGET

Use the formula $a^2 = b^2 + c^2 - 2bc \cos A$ to find the length of a side in a triangle when you are given the lengths of two sides and the included angle.

USING THE COSINE RULE TO FIND THE SIZE OF AN ANGLE

The cosine rule can be used to find the size of an angle given the lengths of the three sides in a triangle. This can be done by changing the subject of the formula.

$$a^2 = b^2 + c^2 - 2bc \cos A$$
$$\Rightarrow 2bc \cos A = b^2 + c^2 - a^2$$
$$\Rightarrow \cos A = \frac{b^2 + c^2 - a^2}{2bc}$$

contd

Therefore, we now have two versions of the cosine rule. We shall refer to this new version as the second version and to $a^2 = b^2 + c^2 - 2bc \cos A$ as the first version from now on.

Again, it is important that you realise that different variables can be used for the second version depending on the given information (see below).

$$\cos A = \frac{b^2 + c^2 - a^2}{2bc} \quad \text{or} \quad \cos B = \frac{a^2 + c^2 - b^2}{2ac} \quad \text{or} \quad \cos C = \frac{a^2 + b^2 - c^2}{2ab}$$

It is unlikely that you will be asked to memorise the formula for the cosine rule, although you probably will memorise it if you use it often enough. In an exam situation, it will appear on the list of formulae. However, be careful when copying down the formula. It is quite common for students to make a copying error when writing it down.

EXAMPLE:

In triangle ABC, AB = 35 metres,
AC = 24 metres and
BC = 31 metres.

Calculate the size of angle BAC.

SOLUTION:

$\cos A = \frac{b^2 + c^2 - a^2}{2bc} = \frac{24^2 + 35^2 - 31^2}{2 \times 24 \times 35}$

$= \frac{840}{1680} = 0.5$

Hence angle BAC = $\cos^{-1} 0.5 = 60°$.

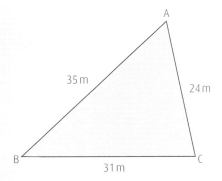

A NON-CALCULATOR EXAMPLE

EXAMPLE:

In triangle PQR, PQ = 7 centimetres,
PR = 5 centimetres and QR = 6 centimetres.

Express the value of cos Q as a fraction
in its simplest form.

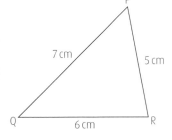

SOLUTION:

$\cos Q = \frac{p^2 + r^2 - q^2}{2pr} = \frac{6^2 + 7^2 - 5^2}{2 \times 6 \times 7} = \frac{36 + 49 - 25}{84} = \frac{60}{84} = \frac{5}{7}$

 THINGS TO DO AND THINK ABOUT

In the four examples below, make a rough sketch of the triangles.

1. In triangle ABC, AB = 9 centimetres, BC = 8 centimetres and AC = 5 centimetres. Calculate the size of angle ABC.

2. In triangle DEF, DE = 10 metres, EF = 11 metres and DF = 19 metres. Calculate the size of angle DEF.

3. In triangle TUV, TU = 12 centimetres, UV = 16 centimetres and TV = 20 centimetres. Calculate the size of angle TUV.

4. In triangle ABC, angle BAC = 45°, AB = 24 centimetres and AC = 29 centimetres. Calculate the length of side BC.

 DON'T FORGET

Use the formula
$\cos A = \frac{b^2 + c^2 - a^2}{2bc}$
to find the size of an angle in a triangle when you are given the lengths of all three sides.

 VIDEO LINK

Watch the clip 'Cosine rule (Finding a Length)' at www.brightredbooks.net/N5Maths

 VIDEO LINK

Watch the clip 'Cosine rule (Finding an Angle)' at www.brightredbooks.net/N5Maths

 ONLINE TEST

Take the test 'The Cosine Rule' online at www.brightredbooks.net/N5Maths

TRIGONOMETRIC PROBLEMS

WHICH FORMULA?

It will save you time if you can quickly identify which formula to use in a problem involving triangles and trigonometry.

Calculating area is straightforward. You can use the formula $A = \frac{1}{2}bh$ in a right-angled triangle *or* when you know the base and height, while you can use the formula $A = \frac{1}{2}ab \sin C$ in other situations. Remember too that Pythagoras' Theorem and basic trigonometry (SOHCAHTOA) can be used in a right-angled triangle.

When calculating the lengths of sides and the sizes of angles using trigonometry, use the sine rule or cosine rule in triangles which are not right-angled. Use the ideas listed below to decide which formula to use.

If you are given 3 sides, use the second version of the cosine rule to find the size of an angle.

If you are given 2 sides and the included angle, use the first version of the cosine rule to find the length of a side.

In all other situations, use the sine rule to find the size of an angle or the length of a side.

DON'T FORGET ➕

Learn how to choose the correct formula quickly, as it will save you a great deal of time and unnecessary effort.

SOLVING PROBLEMS IN TRIANGLES USING TRIGONOMETRY

Many real-life problems can be solved using trigonometry, particularly in areas such as surveying and navigation. You should be aware of the terms 'angle of elevation' and 'angle of depression'.

The **angle of elevation** is the angle an observer must look upwards from the horizontal to see an object.

The **angle of depression** is the angle an observer must look downwards from the horizontal to see an object.

The problems we shall investigate next will involve our triangle formulae, particularly the sine rule and cosine rule; however, they are likely to be two-step problems often requiring the use of the sine rule or cosine rule initially, followed by a second calculation which could involve a second application of a formula or perhaps trigonometry in a right-angled triangle. Watch out in an exam situation for increased marks in a problem. This would indicate that more than one step is required.

EXAMPLE:

Two skyscrapers of equal height are about to be demolished. A surveyor has been asked to calculate their height for reasons of safety.

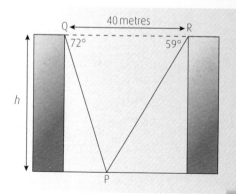

contd

Q and R represent points on the top of the skyscrapers and P represents a point on the ground between them.

The surveyor measures the angle of depression from Q and R to P.

From Q, the angle of depression is 72°, and from R the angle of depression is 59°.

The distance QR is 40 metres.

Calculate the height, h, in metres.

SOLUTION:

In triangle PQR, angle QPR = $(180 - 72 - 59)° = 49°$.

Now apply the sine rule to find PQ (although PR would be suitable as well).

$$\frac{p}{\sin P} = \frac{q}{\sin Q} = \frac{r}{\sin R} \Rightarrow \frac{40}{\sin 49°} = \frac{r}{\sin 59°} \Rightarrow 40 \times \sin 59° = r \times \sin 49°$$

$$\Rightarrow r = \frac{40 \times \sin 59°}{\sin 49°} = 45{\cdot}4.$$

Now focus on the right-angled triangle to the left of the diagram.

Angle P = 72° (alternate angles).

We can use SOHCAHTOA to find h.

$$\sin 72° = \frac{h}{45{\cdot}4} \Rightarrow h = 45{\cdot}4 \times \sin 72° = 43{\cdot}2$$

Hence the height is 43·2 metres.

ONLINE

Find out how surveyors use trigonometry and other mathematical functions in their jobs at www.brightredbooks.net/N5Maths

Advice

This is a typical two-step problem in trigonometry, so study it carefully. Basically, look for a suitable triangle in the diagram, apply an appropriate formula, usually the sine rule or the cosine rule, and complete by using a second triangle.

ONLINE TEST

Take the test 'Trigonometric Problems' online at www.brightredbooks.net/N5Maths

THINGS TO DO AND THINK ABOUT

A class carries out a project to find the height of their school building. The class measures the angle of elevation at two points A and B, as shown below.

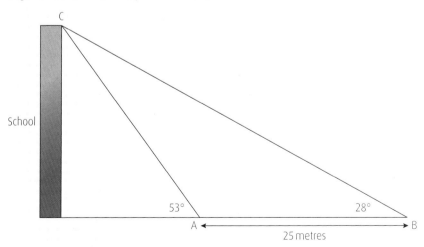

At A, the angle of elevation to C, the top of the school building, is 53°.

At B, the angle of elevation to C, the top of the school building, is 28°.

AB is 25 metres.

Calculate the height of the school.

BEARINGS

THREE-FIGURE BEARINGS

Three-figure bearings are used to describe directions relative to north. The three-figure bearing of north is 000°. The three-figure bearings of other directions are given by angles measured *clockwise from north*. Hence the three-figure bearing of east is 090°, the three-figure bearing of south is 180° and the three-figure bearing of west is 270°.

Many problems on navigation and direction involve bearings and can be solved using trigonometry. It is usually easy to identify the triangle at the heart of any problem, but not so easy to find the sizes of the required angles in the triangle.

EXAMPLE:

A ship leaves port A and sails on a bearing of 071° for 50 kilometres to port B. It then changes course and sails on a bearing of 152° for 63 kilometres to port C. The ship then returns directly to port A.

Calculate the distance AC.

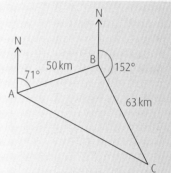

SOLUTION:

The main problem here is that we do not know the size of any angles in triangle ABC. A good tip for helping you to find missing angles is to extend the north line at the vertex B and then look for alternate angles (Z-shapes). You also have the option of extending the line AB and looking for corresponding angles (F-shapes).

We can now find that
angle ABC = (71 + 28)° = 99°
and use the first version of the cosine rule.

$b^2 = a^2 + c^2 - 2ac \cos B$
$= 63^2 + 50^2 - 2 \times 63 \times 50 \times \cos 99°$
$= 7454 \cdot 537$
$= \sqrt{7454 \cdot 537}$
$= 86 \cdot 339\,66$

AC = 86 km (correct to 2 sig. figs).

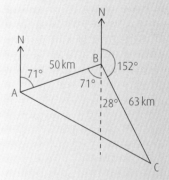

EXAMPLE:

A ship has to deliver supplies to three ports.
Port A is due north of port B.
Port C is on a bearing of 136° from port B.
The distance AB is 200 kilometres.
The distance AC is 420 kilometres.
Calculate the bearing of C from A.

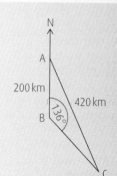

SOLUTION:

We need to find the size of angle BAC, so we must use the sine rule in the triangle to calculate angle ACB first.

contd

$$\frac{a}{\sin A} = \frac{b}{\sin B} = \frac{c}{\sin C} \Rightarrow \frac{420}{\sin 136°} = \frac{200}{\sin C} \Rightarrow 420 \times \sin C = 200 \times \sin 136°$$

$$\Rightarrow \sin C = \frac{200 \times \sin 136°}{420} = 0 \cdot 331$$

Hence angle ACB = $\sin^{-1} 0 \cdot 331$ = 19° (to the nearest degree)

Angle BAC = (180 − 136 − 19)° = 25°

Hence the bearing of C from A = (180 − 25)° = 155° (to the nearest degree).

EXAMPLE:

A ship sails at a speed of 22 kilometres per hour on a bearing of 082° as shown by the dotted line in the diagram. P represents the position of a port. At midday the ship is at Q, which is on a bearing of 310° from P. At 2pm the ship is at R, which is due north of P.

Calculate the distance PQ, correct to 2 sig. figs.

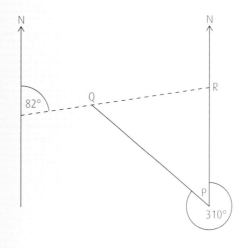

SOLUTION:

Obviously we must use triangle PQR, but we must find information.

QR = 44 km, as the ship sails from Q to R in 2 hours at a speed of 22 km/h.

We can find that angle QPR = 50°. Angle QRP = 82° (alternate angles) and hence as the sum of the angles in a triangle is 180°, then angle PQR = 48°. Now use the sine rule.

$$\frac{p}{\sin P} = \frac{q}{\sin Q} = \frac{r}{\sin R} \Rightarrow \frac{44}{\sin 50°} = \frac{r}{\sin 82°} \Rightarrow 44 \times \sin 82° = r \times \sin 50°$$

$$\Rightarrow r = \frac{44 \times \sin 82°}{\sin 50°} = 56 \cdot 878\,9$$

Hence PQ = 57 km (correct to 2 sig. figs).

 ONLINE TEST

Take the test 'Bearings' online at www.brightredbooks.net/N5Maths

 ## THINGS TO DO AND THINK ABOUT

Finally, here is a tricky problem for you to solve.

An aeroplane flies from airport P to airport Q, and then on to airport R as shown in the diagram.

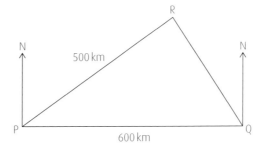

Airport Q is 600 kilometres due east of airport P. Airport R is on a bearing of 325° from airport Q.

The distance from airport P to airport R is 500 kilometres.

Calculate the three-figure bearing of the aeroplane's journey from airport R back to airport P.

TWO-DIMENSIONAL VECTORS

WHAT IS A VECTOR?

A **vector** is often described as being a quantity which has **magnitude** and **direction**. The word 'magnitude' means 'size'. To try to explain this, consider a suitcase being slid along the ground in a straight line. Suppose the case, represented by a yellow rectangle, moves from position A to position B.

Position B

Position A

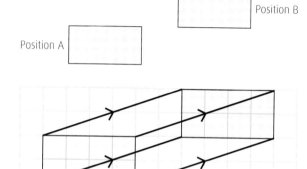

It should be clear that every single point on the case moves the same distance and in the same direction. In the diagram below, a grid and four arrowed lines (at each corner of the suitcase) have been inserted to represent the distance and direction moved by every point on (and in) the case. Can you see that every point on (and in) the case moves (or shifts) 6 units to the right and 2 units up?

The four arrowed lines are representatives of a vector. We say that the **components** of the vector shown are $\binom{6}{2}$, that is, the vector has moved 6 units in the x-direction and then 2 units in the y-direction. Note that the four vectors are *equal*. We could say that only *one* vector is shown in the above diagram, not four.

Vectors are usually represented by a small letter in bold print.

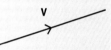

As it is hard to show this when writing on paper, it is common to underline a vector: $\underline{v}$.

NOTE: A quantity which has magnitude but no direction, for example distance, area or temperature, is called a **scalar**.

THE MAGNITUDE OF A VECTOR

The magnitude (or size) of a vector can be found using Pythagoras' Theorem.

EXAMPLE:

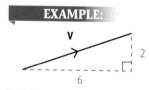

Find the magnitude of vector **v**, shown here.

Give your answer as a surd in its simplest form.

SOLUTION:

The magnitude of $\mathbf{v} = \sqrt{6^2 + 2^2} = \sqrt{36 + 4} = \sqrt{40}$

$\sqrt{40} = \sqrt{4 \times 10} = \sqrt{4} \times \sqrt{10} = 2\sqrt{10}$.

The magnitude of a vector **v** is usually written as $|\mathbf{v}|$. Therefore, it can be said in the above example that $|\mathbf{v}| = 2\sqrt{10}$. This is also referred to as the length of the vector.

In general, the magnitude of a vector with components $\binom{x}{y}$ is $\sqrt{x^2 + y^2}$.

THE COMPONENTS OF A VECTOR

In the earlier example, we saw a vector with components $\binom{6}{2}$. Vectors in two dimensions have two components, a horizontal component in the direction of the x-axis and a vertical component in the direction of the y-axis. The components must be written as a vertical column. Do not confuse components with coordinates such as (6, 2).

contd

EXAMPLE:

Illustrate the following vectors on squared paper.

$\mathbf{a} = \begin{pmatrix} 6 \\ -2 \end{pmatrix}, \quad \mathbf{b} = \begin{pmatrix} -6 \\ 4 \end{pmatrix} \quad \mathbf{c} = \begin{pmatrix} 0 \\ -4 \end{pmatrix} \quad \mathbf{d} = \begin{pmatrix} -4 \\ -4 \end{pmatrix}$

SOLUTION:

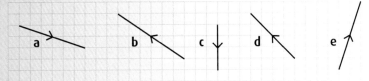

EXAMPLE:

What are the components of vector **e**, shown above?

SOLUTION:

$\mathbf{e} = \begin{pmatrix} 2 \\ 6 \end{pmatrix}$

THE NEGATIVE OF A VECTOR

The negative of a vector is a vector with the same magnitude but opposite direction. We saw in the previous example that $\mathbf{e} = \begin{pmatrix} 2 \\ 6 \end{pmatrix}$. The negative of vector **e** is vector **–e**. Its components are $\begin{pmatrix} -2 \\ -6 \end{pmatrix}$. Both vectors are shown in the diagram alongside.

ADDITION OF VECTORS

We can add vectors in two ways: (i) using a diagram (ii) by adding components.

EXAMPLE:

$\mathbf{u} = \begin{pmatrix} 2 \\ -8 \end{pmatrix}$ and $\mathbf{v} = \begin{pmatrix} 10 \\ 2 \end{pmatrix}$. Illustrate the vector **u** + **v** on a diagram.

SOLUTION:

If we calculate the components of **u** + **v**, we find that $\mathbf{u} + \mathbf{v} = \begin{pmatrix} 2 \\ -8 \end{pmatrix} + \begin{pmatrix} 10 \\ 2 \end{pmatrix} = \begin{pmatrix} 12 \\ -6 \end{pmatrix}$.

Check that the components agree with the diagram.

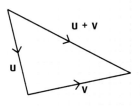

Note what happens if we reverse the order and illustrate **v** + **u** on a diagram. Although the diagram appears different, the resultant vector is the same, in other words **u** + **v** = **v** + **u**. We say that vector addition is commutative, that is, the result is the same even though the order is reversed.

When we add two vectors in the way shown in the above diagram, it is sometimes described as adding them nose to tail. This means that we draw the first vector and start to draw the second vector at the tail (or end) of the first vector.

Earlier, we mentioned the negative of a vector. Can you see that if we add a vector and its negative, for example **e** + (**–e**), we will return to where we started? We say that **e** + (**–e**) = 0, where 0 is called the zero vector.

VIDEO LINK

See a real-life film (from Japan) about a zero vector as a ball is fired backwards out of a truck going forwards: watch 'Vector Addition' at www.brightredbooks.net/N5Maths

THINGS TO DO AND THINK ABOUT

1. Find the magnitude of the following vectors: (a) $\begin{pmatrix} 3 \\ -4 \end{pmatrix}$ (b) $\begin{pmatrix} 5 \\ 1 \end{pmatrix}$.

2. If $\mathbf{u} = \begin{pmatrix} 3 \\ -8 \end{pmatrix}$ and $\mathbf{v} = \begin{pmatrix} 1 \\ 5 \end{pmatrix}$, find $|\mathbf{u} + \mathbf{v}|$.

ONLINE TEST

Take the test 'Two-dimensional Vectors' online at www.brightredbooks.net/N5Maths

MORE VECTORS

SUBTRACTION OF VECTORS

We subtract a vector by adding the negative of the vector. In other words, when we are asked to find **a** − **b**, we can think of it as **a** + (−**b**).

> **EXAMPLE:**
>
> **p** = $\begin{pmatrix} -11 \\ -2 \end{pmatrix}$ and **q** = $\begin{pmatrix} -8 \\ -8 \end{pmatrix}$. Illustrate the vector **p** − **q** on a diagram.
>
> **SOLUTION:**
>
> Think of **p** − **q** as **p** + (−**q**). As **q** = $\begin{pmatrix} -8 \\ -8 \end{pmatrix}$, then −**q** = $\begin{pmatrix} 8 \\ 8 \end{pmatrix}$.
>
> The components of **p** − **q** = **p** + (−**q**), that is **p** − **q** = $\begin{pmatrix} -11 \\ -2 \end{pmatrix} + \begin{pmatrix} 8 \\ 8 \end{pmatrix} = \begin{pmatrix} -3 \\ 6 \end{pmatrix}$.
>
> In the diagram, we use a 'nose to tail' idea by drawing vector **p** first, then start to draw the second vector, −**q**, at the tail (or end) of the first vector.
>
> Check that the components agree with the diagram.

MULTIPLICATION OF A VECTOR BY A SCALAR

> **EXAMPLE:**
>
> Consider the vector **a** = $\begin{pmatrix} 6 \\ 2 \end{pmatrix}$. If we multiply this vector by a scalar (the number 2 in this case), show the vector 2**a** on a diagram.
>
> **SOLUTION:**
>
>
>
> We see that **a** = $\begin{pmatrix} 6 \\ 2 \end{pmatrix}$ and 2**a** = $\begin{pmatrix} 12 \\ 4 \end{pmatrix}$. The vector 2**a** has the same direction as vector **a**, but twice the magnitude.

In general, vectors **a** and k**a** are vectors in the same direction such that the length of vector k**a** is $k \times$ the length of vector **a**.

> **EXAMPLE:**
>
> **u** = $\begin{pmatrix} 5 \\ -2 \end{pmatrix}$ and **v** = $\begin{pmatrix} 1 \\ -3 \end{pmatrix}$. Find the components of (a) 2**u** + 4**v**   (b) **u** − 3**v**.
>
> **SOLUTION:**
>
> (a) 2**u** + 4**v** = $2 \times \begin{pmatrix} 5 \\ -2 \end{pmatrix} + 4 \times \begin{pmatrix} 1 \\ -3 \end{pmatrix} = \begin{pmatrix} 10 \\ -4 \end{pmatrix} + \begin{pmatrix} 4 \\ -12 \end{pmatrix} = \begin{pmatrix} 14 \\ -16 \end{pmatrix}$
>
> (b) **u** − 3**v** = $\begin{pmatrix} 5 \\ -2 \end{pmatrix} - 3 \times \begin{pmatrix} 1 \\ -3 \end{pmatrix} = \begin{pmatrix} 5 \\ -2 \end{pmatrix} + 3 \times \begin{pmatrix} -1 \\ 3 \end{pmatrix} = \begin{pmatrix} 5 \\ -2 \end{pmatrix} + \begin{pmatrix} -3 \\ 9 \end{pmatrix} = \begin{pmatrix} 2 \\ 7 \end{pmatrix}$.

DIRECTED LINE SEGMENTS

Vectors, as we have seen, are represented in a diagram by using lines in which the magnitude of the vector is given by the length of the line, and the direction of the vector is indicated by an arrow. Such representations of vectors are called **directed line segments**. If a vector is used to indicate a shift from one point to another, say from A to B, there is another notation for writing the vector.

contd

The vector **v** shows a shift from A to B. This is also written as $\overrightarrow{AB}$. Therefore we can say that **v** = $\overrightarrow{AB}$. This notation for writing directed line segments shows that the direction of the vector is from A to B. Note that the negative of vector **v** (which has the same magnitude but opposite direction to **v**) can be expressed as −**v** = $\overrightarrow{BA}$; or we can say that −$\overrightarrow{AB}$ = $\overrightarrow{BA}$.

We can add directed line segments in the same way as vectors. In this notation, we could write that $\overrightarrow{AB}$ + $\overrightarrow{BC}$ = $\overrightarrow{AC}$. This would represent a shift from A to C via a third point B.

VIDEO LINK

Go to '6.3 What is a directed line segment or vector?' at www.brightredbooks. net/N5Maths for a short introduction to the basics. Note that the term 'Directional Line Segment' is used in the USA.

POSITION VECTORS

When we are dealing with coordinates of points on a grid in two dimensions, involving an x- and y-axis, we can make use of **position vectors**. A position vector is written in component form and shows how to arrive at a point A starting from the origin.

Hence a point A (x, y) can be represented by a position vector **a** = $\begin{pmatrix} x \\ y \end{pmatrix}$ where **a** = $\overrightarrow{OA}$.

As an example, consider the point A (6, 3). If we start at the origin O and shift to point A, then we must move 6 units in the x-direction and 3 units in the y-direction, therefore point A can be represented by the position vector **a** = $\overrightarrow{OA}$ = $\begin{pmatrix} 6 \\ 3 \end{pmatrix}$. The components of a position vector are the same as the coordinates of the point we shift to from the origin. Remember not to confuse components such as $\begin{pmatrix} 6 \\ 3 \end{pmatrix}$ with coordinates such as (6, 3).

USING POSITION VECTORS

Suppose we are given the coordinates of two points A and B and are asked to find the components of the vector $\overrightarrow{AB}$. While it may be possible to do this by drawing a diagram, the solution can always be found using position vectors. To do this, we must consider a *pathway* of getting from A to B via the origin. This could be written as $\overrightarrow{AB}$ = $\overrightarrow{AO}$ + $\overrightarrow{OB}$. Using this idea, we can find a rule for finding $\overrightarrow{AB}$ using position vectors as shown below.

$\overrightarrow{AB}$ = $\overrightarrow{AO}$ + $\overrightarrow{OB}$ = $-\overrightarrow{OA}$ + $\overrightarrow{OB}$ = $\overrightarrow{OB}$ − $\overrightarrow{OA}$ = **b** − **a**.

The rule $\overrightarrow{AB}$ = **b** − **a** can be used to find the components of any directed line segment if we know the coordinates of points A and B. Study the working and example below to understand this method.

EXAMPLE:

A (7, 3) and B (6, 1) are points on a grid. Find the components of vector $\overrightarrow{AB}$.

SOLUTION:

$\overrightarrow{AB}$ = **b** − **a** = $\begin{pmatrix} 6 \\ 1 \end{pmatrix}$ − $\begin{pmatrix} 7 \\ 3 \end{pmatrix}$ = $\begin{pmatrix} -1 \\ -2 \end{pmatrix}$.
We can check the result by plotting the points on a diagram as shown here.

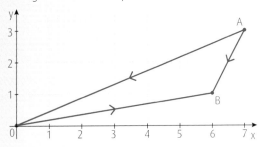

DON'T FORGET

Use the rule $\overrightarrow{AB}$ = **b** − **a** to find the components of a vector given the coordinates of A and B.

ONLINE

Check out 'Position Vectors' at www.brightredbooks.net/ N5Maths

THINGS TO DO AND THINK ABOUT

1. **u** = $\begin{pmatrix} -4 \\ 1 \end{pmatrix}$ and **v** = $\begin{pmatrix} 3 \\ 0 \end{pmatrix}$. Find the components of 4**u** + 6**v**.

2. A, B and C are the points (7, 5), (−4, 2) and (3, −1) respectively.

    Without using a diagram, find the components of the following vectors:

    (i) $\overrightarrow{AB}$ (ii) $\overrightarrow{BA}$ (iii) $\overrightarrow{AC}$ (iv) $\overrightarrow{CA}$ (v) $\overrightarrow{BC}$ (vi) $\overrightarrow{CB}$.

3. Check the results to question 2 using a coordinate grid.

ONLINE TEST

Take the test 'More Vectors' online at www. brightredbooks.net/N5Maths

PROBLEMS INVOLVING VECTORS

PATHWAYS

You will encounter problems on vectors relating to geometric diagrams in which you will have to express a directed line segment in terms of given vectors. This can be done by using pathways through the diagrams. Consider the following examples.

EXAMPLE:

ABCD is a parallelogram as shown here.
M is the midpoint of DC.

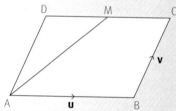

The vector $\overrightarrow{AB}$ is represented by **u**. The vector $\overrightarrow{BC}$ is represented by **v**.

Express (a) $\overrightarrow{DM}$ in terms of **u**;

(b) $\overrightarrow{MA}$ in terms of **u** and **v**.

SOLUTION:

(a) $\overrightarrow{AB} = \overrightarrow{DC} = $ **u** (same magnitude and direction), hence $\overrightarrow{DM} = \frac{1}{2}\overrightarrow{DC} = \frac{1}{2}$**u**.

(b) Find the simplest pathway you can from M to A, namely $\overrightarrow{MA} = \overrightarrow{MD} + \overrightarrow{DA}$.

Hence $\overrightarrow{MA} = -\overrightarrow{DM} - \overrightarrow{AD} = -\frac{1}{2}$**u** $-$ **v**.

> **DON'T FORGET**
>
> Vectors in the same direction as the arrow are positive (+); vectors in the opposite direction to the arrow are negative (−).

EXAMPLE:

In the trapezium, PQ is parallel to SR, and PQ is half the length of SR. $\overrightarrow{SP}$ represents the vector **a**, and $\overrightarrow{PQ}$ represents the vector **b**.

Express (a) the vector represented by $\overrightarrow{RS}$ in terms of **b**;

(b) the vector represented by $\overrightarrow{RQ}$ in terms of **a** and **b**.

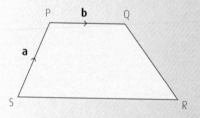

SOLUTION:

(a) $\overrightarrow{RS} = -2\overrightarrow{PQ} = -2$**b**.

(b) $\overrightarrow{RQ} = \overrightarrow{RS} + \overrightarrow{SP} + \overrightarrow{PQ} = -2$**b** $+ $ **a** $+ $ **b** $= $ **a** $-$ **b**.

Other problems on vectors may involve components and magnitude.

EXAMPLE:

The diagram below shows representatives of the vectors **u** and **v**.

(a) Find the components of the vector **u** + **v**.

(b) Find the value of |**u** + **v**|.

(c) Find the components of a vector **w** such that **u** + **w** = **v**.

SOLUTION:

(a) **u** + **v** $= \begin{pmatrix} 10 \\ 4 \end{pmatrix} + \begin{pmatrix} 2 \\ -6 \end{pmatrix} = \begin{pmatrix} 12 \\ -2 \end{pmatrix}$.

(b) |**u** + **v**| $= \sqrt{12^2 + (-2)^2} = \sqrt{144 + 4} = \sqrt{148}$.

(c) **u** + **w** = **v** $\Rightarrow \begin{pmatrix} 10 \\ 4 \end{pmatrix} + $ **w** $= \begin{pmatrix} 2 \\ -6 \end{pmatrix} \Rightarrow $ **w** $= \begin{pmatrix} 2 \\ -6 \end{pmatrix} - \begin{pmatrix} 10 \\ 4 \end{pmatrix} = \begin{pmatrix} -8 \\ -10 \end{pmatrix}$.

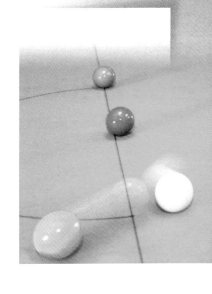

PROBLEMS IN CONTEXT

Vectors appear in many real-life situations: for example, when two snooker balls collide, both change direction at varying speeds; when a boat sails along a river, and the current of the river pulls the boat to the side; and in physics when two forces combine.

EXAMPLE:

An aeroplane travels due east with an air speed of 200 kilometres per hour. At the same time, there is a strong wind blowing due south at 50 kilometres per hour.

Calculate (a) the resultant velocity of the aeroplane;

(b) the actual direction of the flight of the aeroplane.

SOLUTION:

We can represent (i) the direction of the aeroplane (ii) the direction of the wind by vectors and then combine the vectors to show the resultant velocity and direction of the aeroplane.

 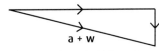

The vector **a** + **w** shows the course taken by the plane as the wind blows it south of its east direction.

(a) We can find the resultant velocity by using Pythagoras' Theorem to find the magnitude of the vector **a** + **w**.

$$|a + w| = \sqrt{200^2 + 50^2} = \sqrt{42\,500} = 206\cdot155$$

Hence the resultant velocity is 206 km/h (to the nearest km/h).

(b) Use trigonometry to find $x°$

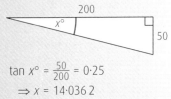

$$\tan x° = \frac{50}{200} = 0\cdot25$$
$$\Rightarrow x = 14\cdot036\,2$$

Hence the actual direction of the plane is 14° south of east (to the nearest degree). This is equivalent to a three-figure bearing of 104°.

NOTE: Speed is a scalar quantity (the rate at which an object covers distance), for example 200 km/h. Velocity is a vector quantity (the rate at which an object changes its position), for example 200 km/h due west.

ONLINE

Check out 'Relative Velocity and Riverboat Problems' at www.brightredbooks.net/ N5Maths

ONLINE TEST

Take the test 'Problems Involving Vectors' online at www.brightredbooks.net/ N5Maths

 THINGS TO DO AND THINK ABOUT

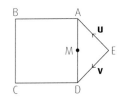

In the diagram, ABCD is a square. $\overrightarrow{EA}$ and $\overrightarrow{ED}$ are the vectors **u** and **v** respectively. Triangle AED is right-angled at E with EA = ED.

(a) Express $\overrightarrow{AD}$ in terms of **u** and **v**.

(b) M is the midpoint of AD. Express $\overrightarrow{AM}$ in terms of **u** and **v**.

(c) Express $\overrightarrow{EB}$ in terms of **u** and **v**.

THREE-DIMENSIONAL VECTORS

COORDINATES IN THREE DIMENSIONS

So far, we have considered vectors in two-dimensional situations only. When coordinates were involved, we needed only an x-axis and a y-axis requiring a two-dimensional grid. When we consider vectors in three dimensions (space), we require a third axis, the z-axis. Think of the x- and y-axes as lying on a horizontal plane and the z-axis as passing through the origin at right angles to the x- and y-axes. Therefore, points in space require three coordinates: an x-coordinate, a y-coordinate and a z-coordinate.

EXAMPLE:

The diagram shows a cuboid OABC, GDEF. Point E has coordinates (5, 3, 4). Write down the coordinates of O, A, B, C, D, F and G.

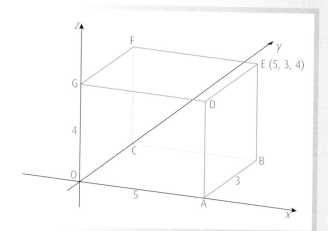

SOLUTION:

You can see from the diagram that to reach E from the origin, you must move 5 units along the x-axis, 3 units along the y-axis and then 4 units along the z-axis, leading to the given coordinates of E (5, 3, 4).

Using this method, the coordinates of the other vertices of the cuboid are:

O (0, 0, 0), A (5, 0, 0), B (5, 3, 0), C (0, 3, 0), D (5, 0, 4), F (0, 3, 4), G (0, 0, 4).

POSITION VECTORS IN THREE DIMENSIONS

All the ideas we covered for vectors in two dimensions apply equally in three dimensions. However, when writing the components of a vector in three dimensions, we now have three components. We can still use the rule $\overrightarrow{AB} = \mathbf{b} - \mathbf{a}$ to find the components of any directed line segment if we know the coordinates of points A and B.

EXAMPLE:

Suppose M is the midpoint of side AD in the earlier diagram of the cuboid OABC, GDEF.

(a) Write down the coordinates of M.

(b) Write down the components of $\overrightarrow{CM}$.

SOLUTION:

(a) M is the point (5, 0, 2).

(b) $\overrightarrow{CM} = \mathbf{m} - \mathbf{c} = \begin{pmatrix} 5 \\ 0 \\ 2 \end{pmatrix} - \begin{pmatrix} 0 \\ 3 \\ 0 \end{pmatrix} = \begin{pmatrix} 5 \\ -3 \\ 2 \end{pmatrix}$.

By looking at the diagram of the cuboid OABC, GDEF, you should be able to confirm that the above result is correct. Consider shifting from C to M by a pathway along the x-direction, then the y-direction and then the z-direction. Therefore $\overrightarrow{CM} = \overrightarrow{CB} + \overrightarrow{BA} + \overrightarrow{AM}$. Check that this satisfies the components $\begin{pmatrix} 5 \\ -3 \\ 2 \end{pmatrix}$.

THE MAGNITUDE OF A VECTOR IN THREE DIMENSIONS

When we studied vectors in two dimensions, we found the following rule for the magnitude of a vector:

The magnitude of a vector with components $\begin{pmatrix} x \\ y \end{pmatrix}$ is $\sqrt{x^2 + y^2}$.

This rule can be extended now to vectors in three dimensions as follows:

The magnitude of a vector with components $\begin{pmatrix} x \\ y \\ z \end{pmatrix}$ is $\sqrt{x^2 + y^2 + z^2}$.

The above ideas are related to using Pythagoras' Theorem in three dimensions (see the example on the flagpole in the section on Pythagoras' Theorem and its converse, page 79).

EXAMPLE:
Find the magnitude of a vector with components $\begin{pmatrix} 6 \\ -8 \\ 24 \end{pmatrix}$.

SOLUTION:
The magnitude is $\sqrt{6^2 + (-8)^2 + 24^2} = \sqrt{36 + 64 + 576} = \sqrt{676} = 26$.

EXAMPLE:
Points A and B have the coordinates (2, –4, 3) and B (5, –2, –6) respectively. Find $|\overrightarrow{AB}|$, the magnitude or length of vector $\overrightarrow{AB}$.

SOLUTION:
First, we must find the components of vector $\overrightarrow{AB}$.
$$\overrightarrow{AB} = \mathbf{b} - \mathbf{a} = \begin{pmatrix} 5 \\ -2 \\ -6 \end{pmatrix} - \begin{pmatrix} 2 \\ -4 \\ 3 \end{pmatrix} = \begin{pmatrix} 3 \\ 2 \\ -9 \end{pmatrix}$$
Hence $|\overrightarrow{AB}| = \sqrt{3^2 + 2^2 + (-9)^2} = \sqrt{9 + 4 + 81} = \sqrt{94}$.

DON'T FORGET

Vectors in two and three dimensions can be treated in exactly the same way. The same rules apply, although vectors in three dimensions have a z-component and the method for finding the magnitude of a vector is extended.

VIDEO LINK

This link has a good clear and visual description of the magnitude of a vector. Some of the terms mentioned are from the next level, but it is still very useful. Go to 'Vectors: Magnitude of a vector 3D' at www.brightredbooks.net/N5Maths

THINGS TO DO AND THINK ABOUT

The diagram shows a cuboid OABC, PQRS

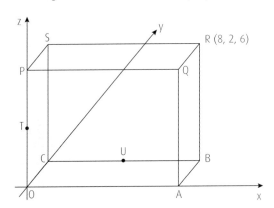

R is the point (8, 2, 6).

T is the midpoint of OP, and U is the midpoint of CB.

(a) State the coordinates of T and U.

(b) Calculate the components of $\overrightarrow{TU}$.

(c) Calculate the magnitude of the vector $\overrightarrow{TU}$.

ONLINE TEST

Take the test 'Three-dimensional Vectors' online at www.brightredbooks.net/N5Maths

USING REVERSE PERCENTAGES

USING A MULTIPLIER

The topic of percentages appears often in real-life situations, from pay rises to examination marks, from interest rates to Value Added Tax (VAT). We met an example earlier in the section on Scientific Notation, where one quantity had to be expressed as a percentage of another. By now, you should be able to carry out basic percentage calculations with and without a calculator. Try out the following example using a calculator.

EXAMPLE:

James works in an office. His current salary is £12 625 per annum. He has been told that next year his salary will increase by 2·4%.

What will his annual salary be next year?

SOLUTION:

Increase in salary = 2·4% of £12 625 = $\frac{2\cdot4}{100}$ × £12 625 = £303

Next year's salary is £12 625 + £303 = £12 928.

We now consider another method of finding the solution to this example. Consider the original salary (£12 625) as 100%. As it will be increased by 2·4%, his new salary will be 100% + 2·4% or 102·4% of his original salary. When using a calculator, we have to change 102·4% to a decimal. As 'per cent' means 'out of 100', we divide 102·4 by 100, leading to 1·024. We can now find next year's salary by multiplying £12 625 by 1·024. Check that this gives the correct solution of £12 928. The number we multiplied by (1·024) is often called the **multiplier**.

EXAMPLE:

Shabaz has bought a car for £10 800. He is told that its value will decrease by 20% after one year.

What will the value of his car be after one year?

SOLUTION:

Consider the original value (£10 800) to be 100%. As it will be decreased by 20%, its new value will be 100% – 20%, or 80% of its original value. As 'per cent' means 'out of 100', we divide 80 by 100, leading to 0·8. We can now find the value after one year by multiplying £10 800 by 0·8. This gives the correct solution of £8640.

It should become fairly simple to find the multiplier with practice. As examples, if we increase a quantity by 3%, the multiplier is 1·03, whereas if we decrease a quantity by 3%, the multiplier is 0·97.

REVERSING A PERCENTAGE CHANGE

Suppose we had taken the previous examples and asked them in a different way. If you had been told that James had received an increase of 2·4% of his salary, that his salary after the increase was £12 928, and you were then asked to find his salary before the increase, how would you do it? In this case, we have to reverse (or undo) the percentage change that has occurred. As we multiplied his original salary by 1·024 to find his new salary, can you see that if you *divide* £12 928 by 1·024, you get the original salary (£12 625)?

We will consider the second example from above and look at how working could be set out.

VIDEO LINK

For more, watch 'Percentage increase 4 – Using Multipliers' at www.brightredbooks.net/N5Maths

DON'T FORGET

Using a multiplier is a very efficient method of calculating values after a percentage increase or decrease has been applied – and we shall use it in the following sections.

contd

EXAMPLE:

In one year, the value of a car decreased by 20%. If the value of the car was £8640 at the end of the year, what was the value of the car at the start of the year?

SOLUTION:

First, find the multiplier. As it was a decrease, subtract 20 from 100, leading to 80%. This in turn gives a multiplier of 0·8 (that is, the value at the start of the year has been multiplied by 0·8 to get £8640). We find the solution by dividing £8640 by 0·8.

Value at start of year = 8640 ÷ 0·8 = £10 800.

You should then check if your solution is correct by doing the original calculation.

EXAMPLE:

Alice buys a laptop. The total cost, including Value Added Tax (VAT) at 20%, was £576.

What was the cost of the laptop before VAT was added?

SOLUTION:

Multiplier = 100% + 20% = 120% = 1·2

Hence cost before VAT = 576 ÷ 1·2 = £480.

To sum up: to reverse a percentage change, add percentage on to 100% if the value has increased, subtract percentage from 100% if the value has decreased, then write down the multiplier. Now divide the new value by the multiplier. You should always check your solution.

Other methods such as proportion are used to teach this process. If you are comfortable using a different method, that is fine.

A COMMON MISTAKE

Many students given the above problem would make the same mistake. They think that as VAT at 20% has been *added* to the original cost, then the answer can be found by *subtracting* 20% from the final cost. So they work out that 20% of £576 is £115·20 and subtract £115·20 to get a final answer of £460·80. Not only is this incorrect as a check will show, but you would get *no marks* if you did this in an examination. This is incorrect because in the example 20% of the value before VAT (£480) was calculated, whereas in the incorrect method, you have worked out 20% of £576.

EXAMPLE:

Joseph has been given an annual pay rise of 2·5%. After the pay rise, his new annual pay is £19 065. What was his annual pay before the pay rise?

SOLUTION:

Multiplier = 100% + 2·5% = 102·5% = 1·025

Salary before pay rise = £19 065 ÷ 1·025 = £18 600.

 THINGS TO DO AND THINK ABOUT

Anne goes for a meal with some friends. The restaurant adds a 15% service charge to the bill. The **total** bill is £48·99.

What was the price of the meal?

 DON'T FORGET

In reverse percentage problems, *never* calculate a percentage and subtract. Study the given examples and practise some more. The correct solution can be found by dividing by the multiplier, assuming you are allowed to use a calculator.

 VIDEO LINK

A short example can be found at 'Reverse Percentage Question 5 Test 1': www.brightredbooks.net/N5Maths

 ONLINE TEST

Take the 'Using Reverse Percentages' test at www.brightredbooks.net/N5Maths

APPRECIATION

GOING UP!

In mathematics, appreciation refers to a quantity going up in value over a period of time. If money deposited in a bank earns interest over a period of time, we say that the money appreciates. We shall consider two methods of finding the solution to problems on appreciation. The first method is popular with many students.

EXAMPLE:

Jonathan buys an antique chair for £1500. It is predicted that its value will increase at the rate of 4·5% per annum.

What will the predicted value of the chair be in 3 years?

Give your answer to the nearest pound.

SOLUTION:

Method 1 – One year at a time

1st year: Increase in value = 4·5% of £1500 = $\frac{4·5}{100}$ × £1500 = £67·50

New value = £1500 + £67·50 = £1567·50.

2nd year: Increase in value = 4·5% of £1567·50 = $\frac{4·5}{100}$ × £1567·50 = £70·5375

New value = £1567·50 + £70·5375 = £1638·0375.

3rd year: Increase in value = 4·5% of £1638·0375 = $\frac{4·5}{100}$ × £1638·0375 = £73·7116875

New value = £1638·0375 + £73·7116875 = £1711·749188.

Hence predicted value after three years is £1712 (to the nearest pound).

SOLUTION:

Method 2 – Using a multiplier

Multiplier = 100% + 4·5% = 104·5% = 1·045

Predicted value after three years = £1500 × 1·045 × 1·045 × 1·045 = £1711·749188.

NOTE: £1500 × 1·045 × 1·045 × 1·045 would normally be written as £1500 × 1·045³.

Hence predicted value after three years is £1712 (to the nearest pound).

DON'T FORGET

You should know how to use your calculator to do 1·045³ without having to do 1·045 × 1·045 × 1·045.

Advantages and Disadvantages

Many students use Method 1, as they feel confident working out percentages and adding. However, it is time-consuming and would be totally unsuitable for longer periods of time than 3 years. The chances of making a mistake increase with each calculation, and early rounding could lead to an incorrect final solution, so all figures must be included throughout the calculation. On the other hand, Method 2 is quick and efficient and avoids the pitfalls of Method 1. There will be occasions when Method 1 may be suitable, but it is essential that you learn, practise and become efficient at Method 2.

EXAMPLE:

The population of Newton is 35 000. It is predicted that the population will rise by 6% per annum. What is the predicted population in 5 years?

Give your answer to the nearest thousand.

SOLUTION:

Predicted population = 35 000 × 1·06⁵ = 46 837·89522

Hence predicted population = 47 000 (to the nearest thousand).

SIMPLE INTEREST

When money is deposited in a bank for a period of less than 1 year, simple interest may be paid. Remember that 'per annum' means 'per year'.

> **EXAMPLE:**
> Find the simple interest paid on £750 invested in a bank for 8 months at 2% per annum.
>
> **SOLUTION:**
> Interest per annum = 2% of £750 = $\frac{2}{100} \times$ £750 = £15
> Interest for 8 months = $\frac{8}{12} \times$ £15 = £15 ÷ 12 × 8 = £10

ONLINE TEST

Take the test 'Appreciation' online at www.brightredbooks.net/N5Maths

VIDEO LINK

Watch the 'Compound Interest' clip prepared by a bank in New Zealand to see the advantages at www.brightredbooks.net/N5Maths

COMPOUND INTEREST

When money is deposited in a bank for periods of longer than 1 year, interest is usually calculated as compound interest.

> **EXAMPLE:**
> Danny has deposited £10 000 in a Platinum Account. It offers a fixed rate of compound interest at 3·1% per annum. How much compound interest will Danny receive over a period of 6 years?
>
> **SOLUTION:**
> Amount in bank after 6 years = £10 000 × 1·031^6 = 12 010·248 45
> Amount = £12 010·25 (to the nearest penny)
> Hence compound interest received = £12 010·25 – £10 000 = £2010·25.

ANNUAL PERCENTAGE RATE (APR)

When borrowing money, many companies charge interest monthly, say 1·125%. If this is then calculated at compound interest, we can find the interest rate for a full year. This is known as the Annual Percentage Rate, or APR.

> **EXAMPLE:**
> A loan company charges its customers 1·125% per month. Find the APR.
>
> **SOLUTION:**
> No matter how much is borrowed, the APR will be the same. First, find the multiplier (1·011 25), then calculate as compound interest for 12 months to make up a full year.
> Hence 1·011 25^{12} = 1·143 674 441.
> This means that the multiplier for a full year is 1·143 674 441. By subtracting 1, we can see that the annual increase, as a decimal, is 0·143 674 441. We multiply by 100 to find the APR, that is 0·143 674 441 × 100 = 14·367 444 1%. Normally, this would be rounded to one decimal place, that is 14·4%.

As you can see, with this company, if you borrow money, the monthly interest rate mounts up to over 14% interest over a full year. It is common for people to borrow money to buy large items such as houses and cars. A loan shark is someone who loans out money, but at a very high rate of interest. When borrowing money, it is important to check that you will be able to pay back the interest and that the APR is not excessively high.

THINGS TO DO AND THINK ABOUT

1. A test on bacteria is being carried out in a hospital. The number of bacteria is increasing at the rate of 0·5% per hour. At 9am, there are 2000 bacteria.
   How many bacteria will there be at 1pm?
   Give your answer correct to 3 sig. figs.

2. A sum of £500 has been deposited in a bank account. How much will be in the bank account after 3 years if the rate of compound interest is 1·75% per annum?

DEPRECIATION

GOING DOWN!

In mathematics, depreciation refers to a quantity going down in value over a period of time. When a car decreases in value over a period of time, we say that its value depreciates. Examples on depreciation are very similar to those on appreciation. We shall use Method 2 from the previous section (using a multiplier). This time, however, the multiplier will be found by subtracting from 100.

EXAMPLE:

A car is valued at £9600. Its value is expected to depreciate by 15% per year.
To the nearest £100, what will the car be worth after 3 years?

SOLUTION:

Multiplier = 100% − 15% = 85% = 0·85
Predicted value after three years = £9600 × 0·85^3 = £5895·6
Hence predicted value after three years is £5900 (to the nearest £100)

EXAMPLE:

A factory uses a piece of machinery which cost £55000 new and will be replaced at the end of the year in which its value falls below half of its value when new. If its value decreases by 20% per annum, when should it be replaced?

SOLUTION:

This is slightly different, as we do not know how many years will be involved, so we have to calculate the value of the machinery a year at a time. We can still use a multiplier, however.

Half of value when new is £55000 ÷ 2 = £27500
Multiplier = 100% − 20% = 80% = 0·8

Value after 1 year = £55000 × 0·8 = £44000
Value after 2 years = £44000 × 0·8 = £35200
Value after 3 years = £35200 × 0·8 = £28160
Value after 4 years = £28160 × 0·8 = £22528

Hence the machinery should be replaced after 4 years, as £22528 < £27500.

> **BEWARE: Some candidates mistakenly would say that the machinery should be replaced at the end of the 3rd year, as 3 × 20% = 60% and 60% > 50%. It does not work like that and must be done as indicated.**

DON'T FORGET

Many questions on appreciation and depreciation ask for the answer to be rounded, and many students forget this final operation. Always check carefully if rounding is required.

GOING UP AND DOWN!

Values may increase and then decrease, or vice versa. For example, many people buy shares in companies as an investment. The value of shares can go up as well as down. One measure is (FTSE) or the Financial Times Stock Exchange 'Footsie' index.

EXAMPLE:

At the start of one week, the FTSE index was 6000.

On Monday it rose by 3·2%, on Tuesday it rose by 1·3%, on Wednesday it fell by 2·1%, on Thursday it fell by 0·2% and on Friday it rose by 1·9%.
What was the FTSE index by the end of the week?

contd

SOLUTION:

This can again be done using multipliers, although in this case there are 5 of them carried out one after the other as shown.

FTSE index = 6000 × 1·032 × 1·013 × 0·979 × 0·998 × 1·019 = 6244·933 386

Hence at the end of the week the FTSE index was 6245.

NOTE: Check carefully all the working, paying particular attention to all five multipliers.

EXAMPLE:

The population of Ashtown is 53 000. The population of Oakville is 40 000. It is predicted that the population of Ashtown will decrease by 5% per annum and the population of Oakville will increase by 4% per annum.

How many years will it take before the population of Ashtown is less than the population of Oakville?

SOLUTION:

After	Ashtown	Oakville
1 year	53 000 × 0·95 = 50 350	40 000 × 1·04 = 41 600
2 years	50 350 × 0·95 = 47 833	41 600 × 1·04 = 43 264
3 years	47 833 × 0·95 = 45 441	43 264 × 1·04 = 44 995
4 years	45 441 × 0·95 = 43 169	44 995 × 1·04 = 46 794

It will be 4 years before the population of Ashtown is less than the population of Oakville.

NON-CALCULATOR PROBLEMS

It would be unusual to be asked to solve a problem on appreciation and depreciation without a calculator. If you were asked, the figures would be fairly easy to manipulate.

EXAMPLE:

The value of a laptop decreased in value from £400 to £320 in one year.

(a) What was the percentage decrease?

(b) If the value of the laptop continues to decrease at this rate, what would the value be after a further year?

SOLUTION:

(a) Actual decrease = £400 – £320 = £80

Percentage decrease = $\frac{80}{400}$ × 100 = $\frac{1}{5}$ × 100 = 20%.

(b) Decrease in value next year = 20% of £320 = £64

Value after further year = £320 – £64 = £256.

Check that you are comfortable doing these calculations without a calculator.

ONLINE

You will find plenty of worked examples to think about and problems to try at 'Appreciation and Depreciation': www.brightredbooks.net/N5Maths

ONLINE TEST

Take the test 'Depreciation' online at www.brightredbooks.net/N5Maths

THINGS TO DO AND THINK ABOUT

A caravan is valued at £15 000. Its value is expected to depreciate by 12% per annum.

What will the caravan be worth after 5 years? Give your answer to the nearest £100.

FRACTIONS

REMINDERS

In several earlier sections, we have had to deal with fractions. You *must* be able to simplify a fraction, for example $\frac{15}{20} = \frac{3}{4}$. You *must* know how to convert a mixed number to an improper (or top-heavy fraction) and vice versa, for example $3\frac{1}{4} = \frac{13}{4}$. You should already be able to add, subtract, multiply and divide simple fractions. We will study how to add, subtract, multiply and divide mixed numbers. This should be done *without* a calculator.

Check the addition and subtraction examples from the earlier section on Algebraic Fractions (page 32) in which we used the Least Common Multiple (LCM) to find a common denominator.

EXAMPLE:

Calculate (a) $\frac{3}{4} + \frac{1}{5}$ (b) $\frac{3}{4} - \frac{1}{5}$.

SOLUTION:

(a) $\frac{3}{4} + \frac{1}{5} = \frac{3 \times 5}{20} + \frac{1 \times 4}{20} = \frac{15}{20} + \frac{4}{20} = \frac{19}{20}$

(b) $\frac{3}{4} - \frac{1}{5} = \frac{3 \times 5}{20} - \frac{1 \times 4}{20} = \frac{15}{20} - \frac{4}{20} = \frac{11}{20}$.

ADDITION OF MIXED NUMBERS

When adding mixed numbers, we add the whole numbers and fractions separately.

EXAMPLE:

Calculate $4\frac{2}{3} + 2\frac{3}{5}$.

SOLUTION:

$4\frac{2}{3} + 2\frac{3}{5} = 6\frac{2 \times 5}{15} + \frac{3 \times 3}{15} = 6\frac{10}{15} + \frac{9}{15} = 6\frac{19}{15} = 6 + 1\frac{4}{15} = 7\frac{4}{15}$.

SUBTRACTION

When subtracting mixed numbers, we subtract the whole numbers and fractions separately.

EXAMPLE:

(a) $5\frac{1}{2} - 1\frac{1}{6}$ (b) $7\frac{1}{4} - 3\frac{2}{3}$.

SOLUTION:

(a) $5\frac{1}{2} - 1\frac{1}{6} = 4\frac{3}{6} - \frac{1}{6} = 4\frac{2}{6} = 4\frac{1}{3}$

(b) $7\frac{1}{4} - 3\frac{2}{3} = 4\frac{1 \times 3}{12} - \frac{2 \times 4}{12} = 4\frac{3}{12} - \frac{8}{12}$ ($\frac{3}{12} - \frac{8}{12}$ is negative)

$= 3 + 1\frac{3}{12} - \frac{8}{12}$ (borrow a whole number)

$= 3 + \frac{12}{12} + \frac{3}{12} - \frac{8}{12} = 3\frac{7}{12}$ (change 1 to $\frac{12}{12}$).

NOTE: In part (b), an issue arose in the numerator when it became $\frac{3}{12} - \frac{8}{12}$. In this type of situation, where the answer is negative, you 'borrow' one of the whole numbers. To do this, we reduce the 4 to 3 and change the whole number borrowed to twelfths $\left(\frac{12}{12}\right)$, as 12 is the denominator in the question. Then 12 is added to the numerators as shown ($12 + 3 - 8$), leading to 7 and the solution.

NOTE: Adding and subtracting mixed numbers can also be done by converting both mixed numbers to improper fractions at the start. This method avoids the issue in part (b) but can lead to large numbers occurring in the calculation, which can be problematic.

MULTIPLICATION

Check the multiplication example from the earlier section on 'More Fractions' (page 34).

EXAMPLE:

Calculate $\frac{2}{3} \times \frac{5}{8}$.

SOLUTION:

$\frac{2}{3} \times \frac{5}{8} = \frac{2 \times 5}{3 \times 8} = \frac{10}{24} = \frac{2 \times 5}{2 \times 12} = \frac{5}{12}$.

To multiply mixed numbers, we must convert the mixed numbers to improper fractions, then multiply the numerators together and the denominators together. If it is possible to cancel (divide the numerator and denominator by the same number), it is often easier if this is done *before* we multiply the numerators together and the denominators together. Otherwise we can end up with large numbers, and it can be difficult to simplify.

EXAMPLE:

(a) $1\frac{3}{5} \times 4$ (b) $2\frac{1}{4} \times 1\frac{2}{3}$.

SOLUTION:

(a) $1\frac{3}{5} \times 4 = \frac{8}{5} \times \frac{4}{1} = \frac{32}{5} = 6\frac{2}{5}$

(b) $2\frac{1}{4} \times 1\frac{2}{3} = \frac{9}{4} \times \frac{5}{3} = \frac{9^3}{4} \times \frac{5}{3_1}$ (cancel by dividing 9 and 3 by 3).

$= \frac{15}{4} = 3\frac{3}{4}$

DIVISION

Check the division example from the earlier section on 'More Fractions' (page 34).

EXAMPLE:

Calculate $\frac{2}{9} \div \frac{1}{3}$.

SOLUTION:

$\frac{2}{9} \div \frac{1}{3} = \frac{2}{9} \times \frac{3}{1} = \frac{2 \times 3}{9 \times 1} = \frac{6}{9} = \frac{2 \times 3}{3 \times 3} = \frac{2}{3}$.

To divide mixed numbers, we must convert the mixed numbers to improper fractions, then leave the first fraction as it is, change the division sign to a multiplication sign, invert the second fraction, that is turn it upside down, then do in the same way as a multiplication of fractions. Again, it is often easier to do any possible cancelling *before* we multiply the numerators together and the denominators together.

EXAMPLE:

Calculate $3\frac{5}{9} \div 1\frac{1}{3}$.

SOLUTION:

$3\frac{5}{9} \div 1\frac{1}{3} = \frac{32}{9} \div \frac{4}{3} = \frac{32}{9} \times \frac{3}{4} = \frac{32^8}{9_3} \times \frac{3^1}{4_1} = \frac{8}{3} = 2\frac{2}{3}$.

THINGS TO DO AND THINK ABOUT

Evaluate the following without a calculator:

(a) $2\frac{2}{3} + 4\frac{2}{3}$ (b) $5\frac{1}{3} - 2\frac{7}{10}$ (c) $3\frac{1}{5} \times 3\frac{3}{4}$ (d) $2\frac{1}{2} \div 4\frac{1}{4}$ (e) $2\frac{1}{4} + \frac{5}{6}$ of $2\frac{4}{5}$.

DON'T FORGET

When multiplying and dividing mixed numbers, it is usually best to cancel first, then multiply, otherwise the numbers can get quite large and it can be difficult to simplify.

VIDEO LINK

Watch a few multiplication and division examples at 'Multiplying and Dividing Mixed Numbers': www.brightredbooks.net/N5Maths

ONLINE TEST

Take the test 'Fractions' online at www.brightredbooks.net/N5Maths

COMPARING DATA SETS

MEASURES OF CENTRAL TENDENCY

We shall be studying statistics in the remaining four sections. When a survey is carried out or an experiment takes place, sets of data are collected. Often, we illustrate the results in a graph or diagram. Another important part of statistics is to analyse the data sets or *distributions* and compare them with other data sets. This can be done by comparing some key measures from the data sets.

There are three measures of central tendency or *averages* which are used when comparing data sets. They are the **mean**, the **mode** and the **median**.

The mean = $\dfrac{\text{Total of all values}}{\text{Number of values}}$;

The mode = the most frequent value;

The median = the middle value in a set of ordered values.

EXAMPLE:

The results of 9 students in a test (out of 20) are listed below.

14	12	18	14	16	10	11	9	13

Calculate (a) the mean;   (b) the mode;   (c) the median.

SOLUTION:

(a) The mean = $\dfrac{\text{Total of all values}}{\text{Number of values}} = \dfrac{14 + 12 + 18 + 14 + 16 + 10 + 11 + 9 + 13}{9} = \dfrac{117}{9} = 13$

(b) The mode = 14

(c) First order the data (lowest to highest) → 9 10 11 12 <u>13</u> 14 14 16 18

The median = 13 (underlined as the middle value).

Note that to find the median was easy, as it was obvious that 13 was the middle value. It is not always so easy, and in fact can become quite difficult when there is a large data set and when there is an even number of values. In the latter case, the median lies in between two values and is found by calculating the mean of these two values.

To help, we have a formula for calculating the position of the median in a list of ordered data.

In a set of ordered data with n values, the position of the median is $(n + 1) \div 2$.

Therefore in the last example, where $n = 9$, the position of the median could be found by using the formula leading to $(9 + 1) \div 2 = 5$. Hence the median was the 5th number in the ordered list.

EXAMPLE:

The percentage marks of 12 students in an exam are listed below.

43	57	62	56	70	38	43	59	80	63	35	91

Calculate the median.

SOLUTION:

Order the data → 35 38 43 43 56 <u>57 59</u> 62 63 70 80 91

As $n = 12$, position of median = $(12 + 1) \div 2 = 6.5$. This means the median is between the 6th and 7th values (underlined). The median is therefore the mean of 57 and 59, which is calculated as $(57 + 59) \div 2 = 116 \div 2 = 58$. Hence the median = 58.

THE RANGE

While measures of central tendency tell us a lot about a data set, they do not tell us everything. Suppose a second group of 9 students had sat the same test (out of 20) as in the first example, and their marks were as follows.

| 12 | 13 | 13 | 13 | 13 | 13 | 13 | 13 | 14 |

If you calculate the mean, you will find that it is also 13, the same as the first group. If we were comparing the two groups, we could say that they had the same mean. However, we can see that the performances of the two groups were very different in other respects. To show this, we have a measure of spread for the data called the **range**.

The range = Highest value – Lowest value

For the two data sets with the same mean, the range is different each time (18 – 9 = 9 for the first set and 14 – 12 = 2 for the second set). The higher the range, the more spread out the data is, so we could conclude that the results in the second set were less spread out or were more consistent than those in the first set.

We use measures of central tendency and measures of spread to compare data sets.

EXAMPLE:

Some exam results are shown in a back-to-back stem-and-leaf diagram.

```
Boys                                                     Girls
            5   2  | 3 | 9
          6 5   1  | 4 | 8
        9 8 6   2  | 5 | 0  4  6  9
          7 5   3  | 6 | 6  6  7  7
            2   0  | 7 | 6  8  9
                4  | 8 | 0  1
n = 15                                        n = 15
                   | 3 | 9     means 39
                 1 | 4 |       means 41
```

(a) Calculate (i) the median for the girls;
    (ii) the range for the girls.

(b) Calculate (i) the median for the boys;
    (ii) the range for the boys.

(c) Compare the performances of the girls and boys, and comment.

SOLUTION:

(a) (i) As $n = 15$, position of median = $(15 + 1) \div 2 = 8$. The results are already ordered, so the median for the girls = 66.

(ii) Range for girls = 81 – 39 = 42.

(b) (i) As $n = 15$, position of median = $(15 + 1) \div 2 = 8$. The results are already ordered, so the median for the boys = 58.

(ii) Range for boys = 84 – 32 = 52.

(c) As the median for the girls is greater (66 > 58), the girls seem to have performed better than the boys. The range for the boys is greater (52 > 42). This suggests that the results for the boys are more spread out.

DON'T FORGET

When comparing data sets, it is not enough to simply compare numbers and say that one is greater/smaller than another. You are expected to reach some kind of conclusion, for example who has done better and whose results are more spread out.

VIDEO LINK

Listen to the 'Mean, Median, Mode song' at www. brightredbooks.net/N5Maths

ONLINE TEST

Take the test 'Comparing Data Sets' online at www. brightredbooks.net/N5Maths

THINGS TO DO AND THINK ABOUT

Find the mean, mode, median and range for the following exam marks.

| 45 | 48 | 38 | 49 | 67 | 51 | 48 | 60 | 54 | 70 |

THE INTERQUARTILE RANGE AND BOXPLOTS

THE QUARTILES

The median divides a data set into two equal halves. The quartiles divide a data set into four equal quarters. To find the quartiles, first find the median of the data set. Then find the median of the first half, called the **lower quartile**. Then find the median of the second half, called the **upper quartile**.

We call the lower quartile Q_1, the median Q_2 and the upper quartile Q_3.

EXAMPLE:

Mrs Smith compares the price of a carton of milk in ten different shops.

£1·09   £1·20   £1·05   £1·43   £1·00   £1·15   £1·25   £1·03   £1·19   £1·36

Calculate (a) the median   (b) the lower quartile   (c) the upper quartile.

SOLUTION:

Order the data → 1·00 1·03 1·05 1·09 <u>1·15 1·19</u> 1·20 1·25 1·36 1·43

As n = 10, position of median = (10 + 1) ÷ 2 = 5·5. This means the median is between the 5th and 6th values (underlined). The median is therefore the mean of 1·15 and 1·19, which is calculated as (1·15 + 1·19) ÷ 2 = 2·34 ÷ 2 = 1·17. Hence the median = £1·17.

The lower quartile is the median of the lower half (1·00 1·03 1·05 1·09 1·15), that is £1·05, and the upper quartile is the median of the upper half (1·19 1·20 1·25 1·36 1·43), that is £1·25.

Hence (a) Q_2 = £1·17   (b) Q_1 = £1·05   (c) Q_3 = £1·25.

THE INTERQUARTILE RANGE

In the previous section, we looked at the range as a measure of the spread of a data set. However, in certain cases, the range can be misleading. Suppose we were calculating the range for the data set 2, 25, 26, 28, 30, 32, 33, 85. The range (85 – 2 = 83) is affected by two values which 'lie outside' most of the other values in the data set. Values such as this are known as **outliers**.

A more useful measure of spread which focuses on the central numbers of the data set is the interquartile range. The interquartile range is the range of the middle half of a data set and is therefore unaffected by any outliers. The formula is given below.

$$\text{The interquartile range} = Q_3 - Q_1$$

In the example on the cost of milk, the interquartile range = £1·25 – £1·05 = £0·20.

Another measure of spread is the semi-interquartile range. It is half of the interquartile range and is also less affected by extreme values. The formula is given below.

$$\text{The semi-interquartile range} = \tfrac{1}{2}(Q_3 - Q_1)$$

EXAMPLE:

The ages of the members of a chess club are listed below.

18   23   56   26   46   23   36   70   32

Calculate the interquartile range.

contd

SOLUTION:

Order the data → 18   23   23   26   32   36   46   56   70

As $n = 9$, position of median = $(9 + 1) \div 2 = 5$, hence the median $(Q_2) = 32$.
The lower quartile is the median of 18, 23, 23, 26. Hence $Q_1 = (23 + 23) \div 2 = 46$
$\div 2 = 23$. This should be obvious without a calculation. The upper quartile is the
median of 36, 46, 56, 70. Hence $Q_3 = (46 + 56) \div 2 = 102 \div 2 = 51$.
Hence the interquartile range = $Q_3 - Q_1 = 51 - 23 = 28$.

BOXPLOTS

A boxplot is a neat visual way of illustrating the key points of a data set. To draw a boxplot,
we need a five-figure summary of the data set. This consists of the lowest value (L), the
three quartiles and the highest value (H).

EXAMPLE:

Mr Ahmed times his car journey, in minutes, from his home to his workplace each
day over a three-week period. The results are listed below.

20   25   18   26   21   35   30   26   25   29   31   18   35   30   27

(a) Calculate (i) median   (ii) the lower quartile   (iii) the upper quartile.

(b) Draw a boxplot to illustrate this data.

Mr Ahmed tries a new route to work over the next three weeks. He times
each journey. The results for the new route are shown in the boxplot below.

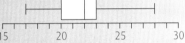

(c) Compare the boxplots and make two appropriate comments about
Mr Ahmed's journey times before and after he changed his route.

SOLUTION:

(a) Order the data → 18 18 20 21 25 25 26 26 27 29 30 30 31 35 35

As $n = 15$, position of median = $(15 + 1) \div 2 = 8$, hence the median $(Q_2) =$
26. Check that Q_1 (the median of 18, 18, 20, 21, 25, 25, 26) = 21 and Q_3
(the median of 27, 29, 30, 30, 31, 35, 35) = 30. Hence the solutions are
(i) 26   (ii) 21   (iii) 30.

(b) The five-figure summary is L = 18, Q_1 = 21, Q_2 = 26, Q_3 = 30, H = 35.

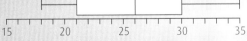

(c) After he changed his route, the journey time was less. We can tell this as the
median fell from 26 minutes to 22 minutes, so the new route seems to be
quicker. The boxplot shows that his journey times are less spread out after he
uses the new route, so the journey times are more consistent.

 DON'T FORGET

If you are asked to draw a
boxplot, use a clear horizontal
scale and make sure that
you use a ruler. It is essential
that boxplots are neatly
drawn. If you are asked to
compare two boxplots, then
you should concentrate
on two things, the median
and the spread, and make
appropriate comments in line
with the advice from the
previous section.

 VIDEO LINK

Watch an excellent example
of a boxplot taking shape
with a comparison to
follow at 'Box Plots': www.
brightredbooks.net/N5Maths

 ONLINE TEST

Take the test 'The
Interquartile Range and
Boxplots' online at www.
brightredbooks.net/N5Maths

 THINGS TO DO AND THINK ABOUT

For the data set: 24 16 17 25 20 34 31 13 28 27 26

(a) Draw a boxplot.

(b) Calculate the semi-interquartile range.

STANDARD DEVIATION

A MORE ACCURATE MEASURE OF SPREAD

We have met three measures which help us to see how a data set is spread out – the range, the interquartile range and the semi-interquartile range. We now consider a more accurate measure of spread called the **standard deviation**. Unlike other measures of spread, the standard deviation uses every member of the data set as part of the calculation. The symbol for standard deviation is s, and there are two formulae which can be used. They are shown below.

$$s = \sqrt{\frac{\Sigma(x - \bar{x})^2}{n - 1}} = \sqrt{\frac{\Sigma x^2 - (\Sigma x)^2/n}{n - 1}}, \text{ where } n \text{ is the sample size.}$$

You will not have to memorise these formulae. However, take great care when copying them. We shall consider an example and use both formulae. Remember that $\bar{x}$ refers to the mean.

EXAMPLE:

A farmer delivers sacks of potatoes to supermarkets in his area. He checks the weight, in kilograms, of a sample of six sacks.

| 48 | 53 | 50 | 51 | 54 | 56 |

Calculate   (a) the mean;   (b) the standard deviation for this data set.

SOLUTION:

(a) Mean = $(48 + 53 + 50 + 51 + 54 + 56) \div 6 = 312 \div 6 = 52$ kg

(b) Method 1

$$s = \sqrt{\frac{\Sigma(x - \bar{x})^2}{n - 1}} = \sqrt{\frac{42}{6 - 1}} = \sqrt{\frac{42}{5}} = \sqrt{8{\cdot}4} = 2{\cdot}9$$
(to 1 decimal place).

x	$x - \bar{x}$	$(x - \bar{x})^2$
48	48 – 52 = –4	16
53	53 – 52 = 1	1
50	50 – 52 = –2	4
51	51 – 52 = –1	1
54	54 – 52 = 2	4
56	56 – 52 = 4	16
		Total = 42

Method 2

$$s = \sqrt{\frac{16\,266 - 312^2 \div 6}{6 - 1}} = \sqrt{\frac{42}{5}} = \sqrt{8{\cdot}4} = 2{\cdot}9$$
(to 1 decimal place).

x	x^2
48	2304
53	2809
50	2500
51	2601
54	2916
56	3136
Total = 312	Total = 16 266

Advice

You will probably prefer *one* of the above methods. Study the layout and the working carefully until you are confident you follow everything. Remember that the symbol Σ is called sigma and means 'the sum of'. Avoid a common mistake if using Method 1. When some students calculate $(x - \bar{x})^2$, they forget that when you square a number, the result is always positive. Many students have difficulty calculating $16\,266 - 312^2 \div 6$ in Method 2. You should do it all in one go on your calculator, write down the answer and *then* divide by the denominator (5 in this case). If you practise these skills, you should become an expert at this.

contd

EXAMPLE:

A sample of six boxes contains the following number of jelly beans per box.

28   27   29   30   32   28

(a) For the above data, calculate:

   (i) the mean;   (ii) the standard deviation.

The manufacturers of the jelly beans claim that 'the mean number of jelly beans per box should be 30 (±2) and the standard deviation should be less than 2'.

(b) Does the data in part (a) support the claim made by the manufacturers? Give reasons for your answer.

SOLUTION:

(a) (i) Mean = $(28 + 27 + 29 + 30 + 32 + 28) \div 6 = 174 \div 6 = 29$

   (ii) Method 1   Method 2

x	$x - \bar{x}$	$(x - \bar{x})^2$
28	$28 - 29 = -1$	1
27	$27 - 29 = -2$	4
29	$29 - 29 = 0$	0
30	$30 - 29 = 1$	1
32	$32 - 29 = 3$	9
28	$28 - 29 = -1$	1
		Total = 16

x	x^2
28	784
27	729
29	841
30	900
32	1024
28	784
Total = 174	Total = 5062

$$S = \sqrt{\frac{\Sigma(x - \bar{x})^2}{n - 1}} = \sqrt{\frac{16}{6 - 1}} = \sqrt{\frac{16}{5}} = \sqrt{3.2} = 1.8 \text{ (to 1 decimal place) or}$$

$$S = \sqrt{\frac{\Sigma x^2 - (\Sigma x)^2/n}{n - 1}} = \sqrt{\frac{5062 - 174^2 \div 6}{6 - 1}} = \sqrt{\frac{16}{5}} = \sqrt{3.2} = 1.8 \text{ (to 1 decimal place).}$$

(b) The manufacturers claim that the mean should be 30 (±2). This means between 28 and 32. The mean is 29, which is between 28 and 32. The standard deviation is 1·8, which is less than 2. So yes, the data does support the manufacturers' claim.

RELATED DATA SETS

EXAMPLE:

A data set has mean 25 and standard deviation 3. Find the mean and standard deviation if:

(a) each member of the original data set is increased by 5;

(b) each member of the original data set is doubled.

SOLUTION:

(a) mean = 30; standard deviation = 3   (b) mean = 50; standard deviation = 6.

NOTE: If we add the same number to each member of a data set, the mean will increase by that amount, but the standard deviation will be unchanged as the spread will remain the same. If we multiply each member of a data set by the same number, then both the mean and standard deviation will be multiplied by that number.

DON'T FORGET

Give detailed answers which show that you understand the question. Answers such as 'Yes' with no explanation are useless. Some students would write 'Yes, because the mean is within 30 (±2) and the standard deviation is less than 2', but this is not enough as it is almost identical to the statement in part (a) of the question. Your answer should mention *your* values for the mean and standard deviation and link them to those made in the claim. Study carefully the solution given and *always* give as full an answer as possible to this type of question.

VIDEO LINK

You can watch a full example of standard deviation being calculated using Method 1 at 'Sample Standard Deviation': www.brightredbooks.net/N5Maths

ONLINE TEST

Take the test 'Standard Deviation' online at www.brightredbooks.net/N5Maths

THINGS TO DO AND THINK ABOUT

Find the mean and standard deviation of the data set 18, 23, 19, 16, 20 and 24.

SCATTERGRAPHS AND PROBABILITY

DRAWING SCATTERGRAPHS

A **scattergraph** is a statistical diagram which is used to compare two data sets. It is used to find connections or a **correlation** between the two data sets. You should already be able to draw a scattergraph, insert a **best-fitting line** and use the line to estimate one value given another.

EXAMPLE:

After a study of taxi fares and cost, the following table is produced.

Journey	A	B	C	D	E	F	G	H
Fare (£)	8	45	18	14	35	26	55	40
Distance (miles)	4	28	12	8	20	16	32	24

(a) Illustrate this data on a scattergraph
(b) Draw a best-fitting line on the scattergraph.
(c) Estimate the distance for a journey with a fare of £50.

SOLUTION:

(a)
(b)

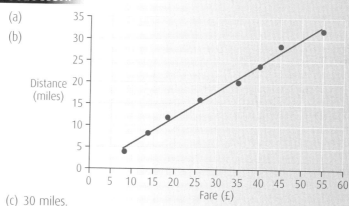

(c) 30 miles.

NOTE: In a scattergraph like the one above, which has a positive gradient, we say there is a positive correlation between the two sets of data: that is, as one increases the other increases.

VIDEO LINK

For more on scattergraphs, watch 'Learn Scatter Plots and Best-Fitting Lines' at www.brightredbooks.net/N5Maths

DON'T FORGET

We can also have a negative correlation between the two sets of data – that is, as one increases the other decreases – and we can have no correlation if the plots on the scattergraph form an unconnected cloud shape.

FINDING THE EQUATION OF A BEST-FITTING LINE

We shall now consider how to find the equation of a best-fitting line. This will be done by thinking back to the section on the equation of a straight line (page 46). We shall use the formula $y = mx + c$ where m is the gradient and c the y-intercept where possible. It would be useful if you did some revision of this topic.

EXAMPLE:

The scattergraph shows the marks of a group of students in their French test (f) and their German test (g). A best-fitting line has been drawn.

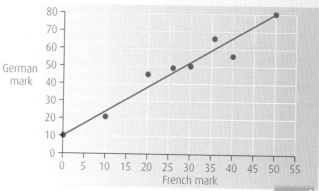

contd

(a) Find the equation of the best-fitting line in terms of f and g.

(b) Use your answer to part (a) to predict the test mark in German for a student who achieved a mark of 60 in French.

SOLUTION:

(a) To find the gradient, select two suitable points on the straight line, for example (0, 10) and (50, 80).

Then find gradient: $m = \frac{y_2 - y_1}{x_2 - x_1} = \frac{80 - 10}{50 - 0} = \frac{70}{50} = 1{\cdot}4$.

As the line crosses the y-axis at (0, 10), the y-intercept is 10, hence $c = 10$.

Using the formula $y = mx + c$, the equation of the straight line is $y = 1{\cdot}4x + 10$.

In terms of f and g, the equation of the best-fitting line is $g = 1{\cdot}4f + 10$.

(b) Substitute $f = 60$ into equation $g = 1{\cdot}4f + 10 \Rightarrow g = 1{\cdot}4 \times 60 + 10 = 94$.

Hence predicted German mark is 94.

 DON'T FORGET

Finding the equation of a best-fitting line is the same as finding the equation of any straight line, so it is very important that you are confident at doing this.

PROBABILITY

Probability is a measure of the chance of an event happening. If an event is *impossible*, its probability is 0. If an event is *certain*, its probability is 1. The probability of any other event happening is somewhere between 0 and 1. Probability is defined as

$$\frac{\text{Number of favourable outcomes}}{\text{Total number of outcomes}}.$$

EXAMPLE:

The table shows the results of a survey of fourth-year pupils.

	Wearing school uniform	Not wearing school uniform
Girls	48	12
Boys	37	23

What is the probability that a fourth-year pupil, chosen at random from this sample, will be a girl wearing school uniform?

Express your answer as a fraction in its simplest form.

SOLUTION:

$$\text{Probability} = \frac{\text{Number of favourable outcomes}}{\text{Total number of outcomes}} = \frac{48}{120} = \frac{24}{60} = \frac{12}{30} = \frac{2}{5}$$

 ONLINE TEST

Take the test 'Scattergraphs and Probability' online at www.brightredbooks.net/N5Maths

 THINGS TO DO AND THINK ABOUT

The graph shows a scattergraph with a best-fitting line inserted.

(a) Find the equation of the best-fitting line.

(b) Use your answer to part (a) to predict y when $x = 30$.

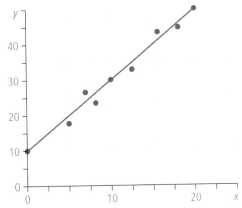

amplitude
stretches a graph vertically, for example 3 in $y = 3 \sin x°$.

angle of depression
the angle at which you look down from the horizontal to see an object.

angle of elevation
the angle at which you look up from the horizontal to see an object.

arc
part of the circumference of a circle.

axis of symmetry
a line of symmetry on a graph such that the two sides on either side of the line are mirror images of each other.

best-fitting line
drawn on a scattergraph to follow the pattern of the dots.

components
written vertically to show the magnitude and direction of a vector.

converse of Pythagoras' Theorem
states that 'in a triangle, if $a^2 = b^2 + c^2$, then angle A is a right angle'.

correlation
tells us how two variables such as height and weight are related.

cosine rule
$a^2 = b^2 + c^2 - 2bc \cos A$ or $\cos A = \frac{b^2 + c^2 - a^2}{2bc}$; used for solving trigonometric problems in triangles.

cube root
a number that must be multiplied by itself three times to equal a given number, for example $\sqrt[3]{1000} = 10$.

directed line segment
a way of representing a vector, for example $\overrightarrow{AB}$.

discriminant
$b^2 - 4ac$: tells you the nature of the roots of a quadratic equation.

factorisation
the opposite process to multiplying out brackets.

function
a relationship between two sets such that each member of the first set is related to exactly one member of the second set, for example $f(x) = 2x$.

gradient
tells you the slope of a straight line.

gradient formula
given by $m = \frac{y_2 - y_1}{x_2 - x_1}$.

hypotenuse
the side opposite the right angle in a right-angled triangle.

index (plural: indices)
another name for a power.

inequation
a sentence containing 'greater than' or 'less than' symbols.

interquartile range
a measure of spread given by $Q_3 - Q_1$.

irrational numbers
numbers which cannot be represented by a fraction, for example π and $\sqrt{2}$.

justify
means that you are expected to explain in words the solution to a problem.

kite
a quadrilateral with two pairs of equal adjacent sides.

Least Common Multiple (LCM)
the smallest multiple which is common to two or more numbers.

lower quartile
denoted by Q_1; the smallest of the three quartiles which divide a data set into four equal parts.

magnitude
the length or size of a vector.

maximum turning point
the type of turning point on a ∩-shaped parabola.

mean, mode and median
three measures of central tendency used to analyse data sets.

minimum turning point
the type of turning point on a ∪-shaped parabola.

multiplier
the number we multiply by in order to carry out a percentage increase or decrease.

octagon
an 8-sided polygon.

parabola
the graph of a quadratic function.

perfect square
any number which is the square of a rational number, for example 0, 1, 4, 9, 25 etc.

perimeter
the distance around the outside of a shape.

quadratic expression
an expression of the form $ax^2 + bx + c$, where $a \neq 0$.

the quadratic formula
$x = \frac{-b \pm \sqrt{(b^2 - 4ac)}}{2a}$ is used to find the roots of a quadratic equation, especially when the roots are irrational numbers.

rational numbers
denoted by Q and can be expressed as a fraction.

real numbers
denoted by R; the set of all rational and irrational numbers.

scalar
a quantity with magnitude but no direction, for example temperature.

scale factor
the ratio of any two corresponding sides in two similar shapes.

scattergraph
a statistical diagram used to compare two data sets.

scientific notation or **standard form**
a method of writing very large or very small numbers in a more compact form.

semi-interquartile range
a measure of spread given by $\frac{1}{2}(Q_3 - Q_1)$.

standard deviation
a measure of spread given by
$s = \sqrt{\frac{\Sigma(x - \bar{x})^2}{n - 1}} = \sqrt{\frac{\Sigma x^2 - (\Sigma x)^2/n}{n - 1}}$, where n is the sample size.

tangent
a straight line which touches a circle at one point only or a trigonometric ratio found by dividing the opposite side by the adjacent side in a right-angled triangle.

tangent-kite
a quadrilateral whose sides are two tangents drawn to a circle from a point outside the circle, and two radii.

upper quartile
denoted by Q_3; the largest of the three quartiles which divide a data set into four equal parts.

vertex (plural: vertices)
the point at the corner of an angle also used for the turning point of a parabola.

vertical translation
slides a graph up or down in the direction of the y-axis.

wave
term often used to describe the shape of trigonometric graphs such as the sine graph.

Z
the set of integers $\{... -3, -2, -1, 0, 1, 2, 3 ...\}$, the set of positive and negative whole numbers and zero, named after the first letter of the German word *Zahlen*, meaning numbers.